China's Journey to Space:
From Dream to Reality

By: Xu Jing

China Intercontinental Press

图书在版编目（CIP）数据

梦圆太空：中国的航天之路：英文 / 徐菁著；郑慧贞译.
-- 北京：五洲传播出版社，2013.12
ISBN 978-7-5085-2702-4
Ⅰ．①梦… Ⅱ．①徐… ②郑… Ⅲ．①航天工业－发展史－中国－英文 Ⅳ．① F426.5
中国版本图书馆CIP数据核字（2013）第292271号

"中国创造"系列

策　　划	荆孝敏　付　平
主　　编	付　平
出 版 人	荆孝敏

梦圆太空——中国的航天之路

著　　者	徐　菁
图片提供	神舟传媒　秦宪安　宿　东　李　刚　刘淮宇　南　勇等
责任编辑	黄金敏
助理编辑	姜　超
翻　　译	郑慧贞
审　　校	Cavan Reilly（英国）
装帧设计	张　凭　李洪涛
制　　作	北京新影响文化发展有限公司
出版发行	五洲传播出版社
	（北京市海淀区北三环中路31号凯奇大厦）
承 印 者	北京全海印刷有限公司
版　　次	2014年4月第1版第1次印刷
开　　本	710mm*1000mm　16
印　　张	13
字　　数	70千
印　　数	1-3000册
定　　价	98.00元

Contents

Foreword 1

Milestones in the Development of China's Space Industry 2

1. Initiation of China's Aerospace 5
2. Exploration of China's Manned Spaceflight 20
3. A New Journey for Lunar Exploration 31

Long March Rocket Family 38

1. LM-1 Carrier Rocket Series: China's First Carrier Rocket Models 41
2. LM-2 Carrier Rocket Series: Sending Shenzhou Spacecraft into Space 42
3. LM-3 Carrier Rocket Series: Ladder to the Moon 47
4. LM-4 Carrier Rocket Series: Sending Satellites into Sun Synchronous Orbit 55

Chinese Satellites in Space 58

1. Widely-used Earth Observation Satellites 60
2. Communications Satellites as Bridges for Space Information 76
3. Beidou Navigation Satellites 81
4. Experimental Satellites for Scientific and Technological Development 86
5. Chinese Satellites in Global Market 90

The Shenzhou Spacecraft: Chinese Nation's Millennium Dream of Spaceflight Comes True   92

1. China's Self-developed Flying Vehicle: The Shenzhou Spacecraft   95

2. Shenzhou 5 Mission: First Manned Spaceflight   97

3. Shenzhou 6 Mission: First Multi-manned and Multi-day Spaceflight   100

4. Shenzhou 7 Mission: First Spacewalk   102

5. Shenzhou 8 Mission: First Rendezvous and Docking in Space   106

6. Shenzhou 9 Mission: Travel of First Female Astronaut to Space   110

7. Shenzhou 10 Mission: First Applied Spaceflight   115

Travel of Chinese Astronauts to Space   122

1. "Cradle" of Chinese Astronauts   124

2. Selection and Training of Chinese Astronauts   126

3. Unusual Space Life of Chinese Astronauts   132

4. "Feitian" Space Suit: Special Space Suit for Chinese Astronauts   140

5. Chinese Space Food   142

The Wonderful Beginnings and Progress of the Chang'e Program   144

1. First Lunar Exploration by Chang'e 1   147

2. Second Step in Lunar Exploration by Chang'e 2   159

3. First Landing on the Moon by Chang'e 3   166

Desert, Mountain and Ocean: Spaceports for China's Satellite Launch   172

1. First Spaceport for Manned Spaceflight: Jiuquan Satellite Launch Center   174
2. Aerospace Town on the Plateau in Northwestern Shanxi: Taiyuan Satellite Launch Center   182
3. "Moon City" Helping Chang'e with the Journey to the Moon: Xichang Satellite Launch Center   184
4. Future Manned Lunar Landing Spaceport: Wenchang Satellite Launch Center   190

The Broad Prospects of China's Aerospace Sector   192

1. New Carrier Rockets   194
2. Various Artificial Earth Satellites   196
3. Future Manned Space Station   198
4. Deep Space Exploration in Progress   199

Foreword

At 1:30, early in the morning of December 2nd, 2013, the "Chang'e 3" lunar probe flew to the Moon with the "Yutu" lunar rover on board, which realized China's first soft landing onto the Moon. The exploration of outer space has long been the relentless pursuit of mankind: In 1961, Yuri Alekseyevich Gagarin, cosmonaut of the former Soviet Union, spent the most precious 89 minutes of his life in orbit of the Earth, starting the journey of mankind into space; In 1969, American astronaut Neil Alden Armstrong took the famous "small step" on the surface of the Moon, marking mankind's setting foot on a land other than the Earth for the first time…

China's space industry started from the 1950s and 1960s, and the successful launching of China's first man-made earth satellite Dongfanghong 1 in 1970 ushered China into the space era.

Through more than five decades' efforts, China's space industry has probed an innovative approach independently which is suitable for Chinese national circumstances. The manufacturing and launching of carrier rockets, artificial earth satellites of all kinds, manned spacecraft and lunar probes are all milestones which represent leapfrog development in China's space industry.

Chinese poet Li Bai had written such beautiful verses: "There must be a day when I will ride on winds and waves, set my cloud-white sail and cross the sea to the shore of ideals." By seizing the opportunities brought by the "Twelfth Five-year Plan" (2011-2015), China's space industry will stand at a new start point for fulfilling new space dreams.

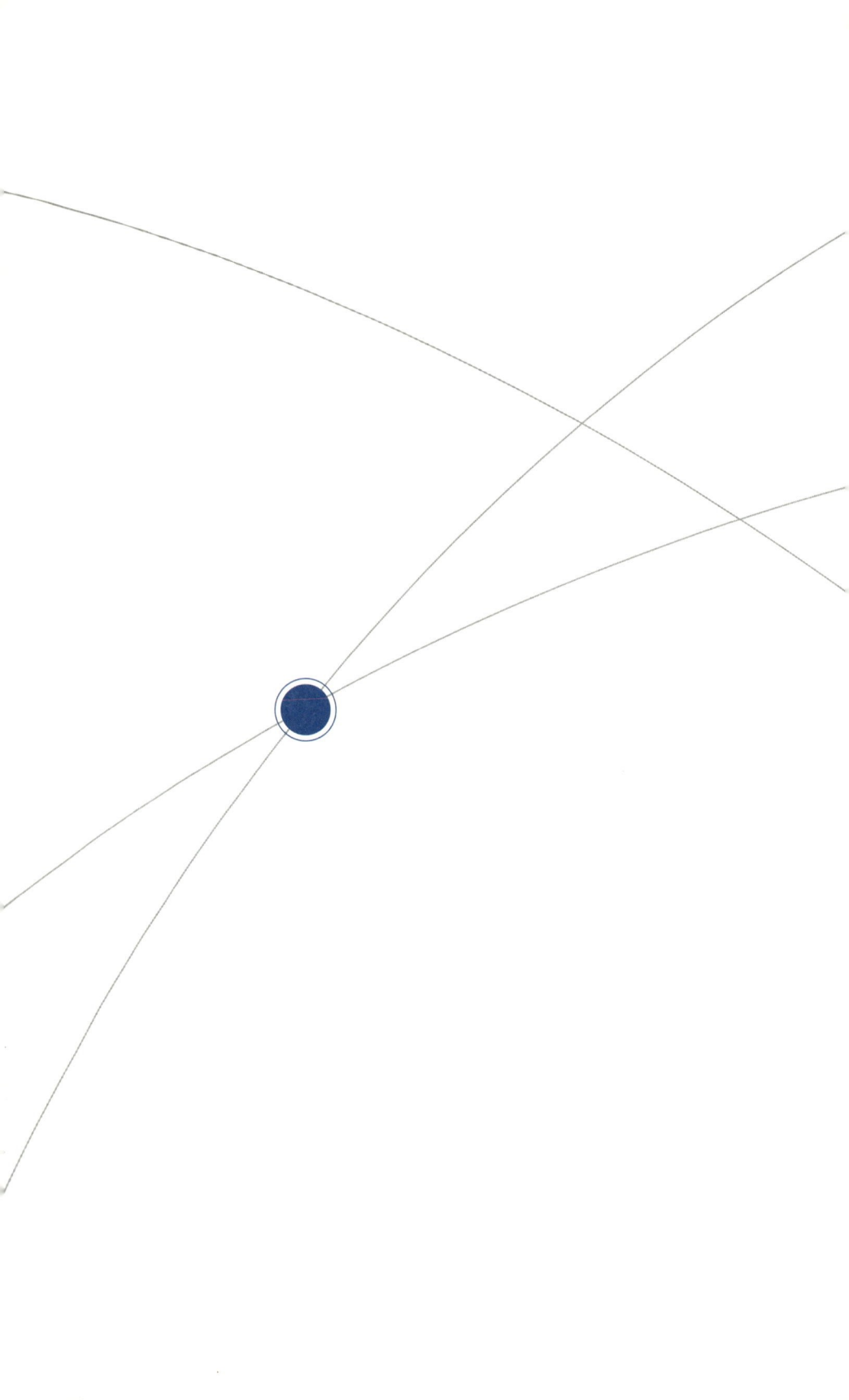

1

Milestones in the Development of China's Space Industry

China's Journey to Space: From Dream to Reality

The development of China's space industry is marked by three milestones:

On April 24th, 1970, China's first artificial earth satellite, Dongfanghong 1 (DFH-1, "Dongfanghong" literally means "the east is red"), was successfully launched and operated in orbit, ushering in a new era in the development of China's space industry.

On October 15th and 16th, 2003, China achieved great success in the launch of its first manned spacecraft, Shenzhou 5, into space, becoming the third nation in the world to develop the capability of independently carrying out manned spaceflight programs.

On October 24th, 2007, China's first lunar exploration satellite, Chang'e 1 (named after the goddess Chang'e who reached the Moon in an ancient Chinese fairy tale), flew to the Moon and from November 20th began to send back clear Moon images, marking that China had become one of the few countries in the world with deep space exploration capabilities.

| Milestones in the Development of China's Space Industry |

Initiation of China's Aerospace

"When the east is red, the sun rises…" When the music "The East Is Red" was broadcast in space, it marked that the Chinese nation had taken the first step towards flight into space and also let us remember the glory days.

▶ Proposal for the Man-made Satellite Program

At the end of 1956, Nie Rongzhen was made Vice Premier of the State Council to take charge of national science and technology work. In the face of the challenges of the world today and of the future, and despite illness, this battle-scarred general went on to lead China's scientific and technological force.

In 1957, the first man-made satellite in the world was sent into space. Coordinated by the Chinese Academy of Sciences (CAS), Qian Xuesen, Zhao Jiuzhang, Guo Yonghuai, Lu Yuanjiu and other experts developed the draft for the man-made satellite development program with the "three-step" development idea: first, to launch sounding rockets; second, to place satellites weighing 100-200 kg into space; and third, to send satellites weighing thousands of kilograms into space.

CAS made man-made satellite development its first priority among

▶ Former Site of 581 Group

scientific and technological tasks for 1958, for which China's first satellite group, "CAS 581 group", was set up with Qian Xuesen as the head and Zhao Jiuzhang and Wei Yiqing as associate heads. Under their leadership, three design institutes were established: the first design institute was engaged in overall design of satellites and carrier rockets;

▶ Ground Tracking Station

▶ Upper Stage of LM-1 Rocket

the second institute was dedicated to the development of the control system; and the third institute was committed to the development of space exploration instruments and research on the environment in space.

At that time, due to extremely limited economic support and technical strength, China had to give priority to the development of missiles and atomic energy for the sake of national defense. In terms of technical strength, China had just begun to pattern short-range missiles and had not developed the ability to independently design carrier rockets. Only after China's rocket technology achieved further development could China develop large rockets able to carry man-made satellites. Under these circumstances, studies were started looking into individual technologies for the man-made satellite program.

The newly established satellite group had to start from scratch, empty-handed. Without an office, they rented several rooms at Xiyuan Hotel in the western suburbs of Beijing. Without advanced computers, they used hand-operated computers. Without desks, they developed the design drawings on the cement floor. With a set of pliers, two files, several aluminum sheets and ternary plates and a dozen candles and several flashlights, they started to design and develop the prototypes of Chinese satellites and carrier rockets.

On the National Day in 1958, the Exhibition for Leapfrog Achievements of CAS in Natural Science had its curtain raised at the Beijing Zhongguancun Biology Research Institute. For the exhibition, the satellite group rushed out a set of carrier rocket design drawings, ground radar photos and satellite and rocket models. As a result, the exhibition created much of a stir, not only receiving recognition and encouragement from all walks of life, but also attracting the attention of the central leaders.

In 1961, several scientists including Qian Xuesen and Zhao Jiuzhang decided to pool together scientists from various disciplines by organizing occasional interplanetary travel forums so that they could first figure out the theoretical side of the relevant space technology problems and, thus, lay a solid theoretical foundation for launch of man-made satellites in the next phase.

▶ Docking of DFH-1 with Upper Stage of Rocket

Qian Xuesen's speech at the forum was entitled "The current state of rocket power for interplanetary travel in the Soviet Union and the United States and its prospects". The speech lasted more than an hour and Qian Xuesen finished it in one breath. Throughout Qian's speech, each expert and scholar present was highly excited. After Qian finished his speech, Pei Lisheng, Zhu Kezhen, Bei Shizhang, Zhao Jiuzhang, Guo Yonghuai and other scientists expressed their opinions, which were previously borne in silence, one after another, enlivening the academic atmosphere of the forum...

The first interplanetary travel forum aroused an enthusiastic response from the scientific community. It was widely believed that forums of this kind were of great significance and should continue to be held from time to time in the future. Accordingly, in the three years that followed, CAS convened 12 similar interplanetary travel forums in total. At each

▶ Refueling

▶ DFH-1 Satellite under Test

of the forums, an expert made a keynote speech, which was followed by a discussion about this speech with the attendees where they could fully put forward their opinions and suggestions. After each forum, the opinions expressed were summarized and made into articles, with 200 copies printed and issued to relevant departments. Finally, the papers read out at the forum were also collected into the "Collection of Interplanetary Travel Materials" published by the Science Press, and received wide spread attention from the academic community and other walks of life. Consequently, for the study and discussion of China's man-made satellite technology issues, a preliminary leadership and a structured plan for an organization began to take shape and achieved remarkable results, creating the circumstances necessary for the future creation of the man-made satellite program.

▶ Music Device on DFH-1

June 29th, 1964 saw the successful launch of China's first self-developed ballistic missile, and October 16th witnessed the successful detonation of China's first atomic bomb. Successful development of the missile and the atomic bomb laid the technical foundations and provided a guarantee of the conditions needed for the launch of man-made satellites. Ultimately, the development of man-made satellites was now on the agenda.

At the end of 1964, Premier Zhou Enlai received a letter from Zhao Jiuzhang regarding the commencement of the development of man-made satellites. At the same time, after a thorough analysis of the points in favor for China's development of man-made satellites, Qian Xuesen, advocate of China's space cause, also put together a man-made satellite development plan. In April, 1965, the National Defense Science and Technology Commission drafted "The Report on the Man-made Satellite Program". After approval from the central authorities, the decision in favor of developing man-made satellites had finally been made. This marked the official commencement of China's man-made satellite program for design and organizational research.

▶ Satellite Assembled into Upper Stage of Rocket

▶ Assembly of DFH-1 Satellite

Late during one night in April 1965, satellite expert Pan Houren received a call out of the blue from Zhao Jiuzhang. When Pan rushed to Zhao's home, Guan Zhaozhi, Head of CAS's Institute of Mathematics, was already there. After a brief introduction to each other, Zhao Jiuzhang said eagerly: "At the end of last year, I wrote to Premier Zhou, explaining that we had already met the requirements for developing man-made earth satellites and, for the purpose of missile development, suggested a combination of missile shooting tests and a satellite launch, so as to achieve double gains." After pausing for a while, Zhao turned pages of a little notebook in his hand and continued excitedly: "Now Premier Zhou has instructed us to work out a preliminary plan. Since 1958, we have been making preparations and looking forward to this day, and now it's coming at long last. However, just imagine if a satellite, several meters long, is sent into orbit and we can't control it and we end up leaving it floating like a balloon several kilometers away from us. So before we launch the satellite, we must set out satellite movement rules, orbit calculation, surveillance, control and forecast and tracking station layout. In this respect, CAS ought to take responsibility for the mission

| Milestones in the Development of China's Space Industry |

and be a step ahead of others. So I hope Mr. Guan can coordinate the relevant personnel and implement the mission as soon as possible."

Zhao Jiuzhang also said to Pan Houren: "Since you are an expert on astronomy and have previously worked in the field of satellite orbit, I hope you can participate in this program on behalf of the '581' group and work and coordinate with the Institute of Mathematics." On the next day, the general satellite group was established, which was made up of Pan Houren, He Zhenghua and Hu Qizheng and chaired by He Zhenghua. At the same time, coordinated by Guan Zhaozhi, the mechanics group headed by Ye Shuwu was also compelled to be engaged in satellite orbit calculation, and personnel at the Purple Hills Observatory were also invited to participate in the program. In this way, CAS pooled together the strongest satellite development groups and set up a joint group dedicated to satellite orbit calculation, which immediately went to work on orbit determination, calculation and forecast after satellite launch. Meanwhile, other relevant institutes under CAS also went into action.

In August 1965, CAS submitted to the central authorities the report

▶ Pep Rally before Launch

China's Journey to Space: From Dream to Reality

entitled "Suggestions for the Development of China's Man-made Satellite Program", which put forward guiding principles for the development of China's space technologies. In August, the 13th session of the Central Special Committee presided over by Premier Zhou Enlai approved the report in principle, identified the development of man-made satellites as a significant task in the development of national sophisticated technologies and designated the National Defense Science and Technology Commission as the organization and coordination unit for the satellite program. As the suggestions were proposed in January 1965, the program code for China's first man-made satellite was "651".

▶ Jiuquan Launch Site

▶ Development of the Blueprint

In September 1965, CAS formally set up the Satellite Design Institute to be responsible for satellite design, development and overall coordination and set about putting together the plan for China's first man-made satellite. Qian Ji who took charge of the overall satellite group reported to Premier Zhou Enlai. When Premier Zhou found out that Qian Ji was also surnamed Qian, he said humorously: "Our chief satellite designer [Qian Xuesen] is also surnamed Qian. Cutting-edge technology work, like the development of missiles, atomic bombs and satellites, cannot do without "Qian" (referring not only to those scientists surnamed Qian and engaged in high technology, but also to money, the literal meaning of qian). (Note: Qian Xuesen was engaged in missile development; Qian Sanqiang was committed to atomic bomb development; and Qian Ji was dedicated to satellite development.)

From October 20th to November 30th, 1965, the meeting on the feasibility of the overall program for China's first man-made satellite was held in Beijing. This meeting was unprecedented for its long duration, large scale and rich contents. Many experts recalled later that this had been the longest meeting they had ever been to in their life. It was determined at this meeting that China's first man-made satellite should be experimental and for scientific exploration, aimed at laying a technical and practical foundation for China's development of earth observation satellites, communications and broadcast satellites, meteorological satellites and other applied satellites.

What signals should China's first man-made satellite transmit? According to experts at the meeting, China's first man-made satellite should transmit continuous signals which not only had Chinese characteristics but would also be recognized by the world. For this, China National Radio recommended adopting the melody of "Dongfanghong". Since the tune of "Dongfanghong" had already became a symbol of "Red China", it was suggested that the first satellite should broadcast "Dongfanghong" and that the satellite be named after the melody, i.e. "Dongfanghong 1".

On December 11th, 1967, the National Defense Science and Technology Commission convened a meeting on the implementation of the satellite development program. At the meeting, the commission simplified the satellite program, officially named China's first satellite "Dongfanghong 1" (DFH-1) and also set the overall technical objectives for DFH-1, namely: "to be able to lift off, to catch it, to hear it and to see it."

"Able to lift off" meant the successful launch of the satellite; "able to catch it" indicated that the satellite should be able to be tracked, measured and controlled in a remote manner; "able to hear it" meant that the content broadcast by the satellite should be heard; and "able to see it" implied that the satellite should be able to be seen with telescopes or naked eyes. Among the four objectives, "able to lift off and catch it" were crucial to the satellite program, and "able to hear and see it" were intended to enlarge the influence of the satellite. As for the objective of being "able to hear it", engineers and technicians finally adopted compound tones produced by electronic circuits to simulate an aluminum piano. By playing music, high stability sound-source oscillators act as voice keys and tempos generated by the program control circuit are used to control articulation of the sound source oscillators, achieving satisfactory music effects.

As to how to "see" China's satellite from the Earth and with only the naked eye, engineers and technicians also went through painstaking efforts, because it is far beyond the usual realms of possibility to see a satellite weighing 173 kg and measuring a maximum of 1 m in diameter after it is sent into space. In order to see the satellite, engineers and technicians installed an "observation ball" on the third stage of the Long March 1 (LM-1) carrier rocket that would go into space together with the satellite. The observation ball was made of reflecting materials and characterized by having a large volume and a light weight. After it was brought into space, it opened and flew by attaching itself onto the satellite. When seen from the Earth, it looked like a bright star.

▶ The Birth of China's First Satellite

In February 1968, the China Academy of Space Technology (CAST) was officially established with Qian Xuesen as the president and Chang Yong as the political commissar. The academy pooled together the research strength of all departments to concentrate on the development of DFH-1. The head of the overall satellite design department was 37-year-old Sun Jiadong. Later, Sun Jiadong recalled that when he learned that his appointment had been proposed by Qian Xuesen and approved by Nie Rongzhen, he was very excited, but also felt an increased pressure at the same time.

The young Sun Jiadong acted with enormous courage and determination and demonstrated great tenacity. After a thorough study and analysis of the satellite research and development conditions at that time, he spent more than two months investigating those with special technical expertise across various departments and finally selected the 18 most promising candidates, including Qi Faren, the celebrated academician from the Chinese Academy of Engineering (CAE), who had participated in a wide range of space programs ranging from DFH-1 to the Shenzhou spacecraft. These 18 technical experts were later known as the "18 space warriors".

▶ Chairman Mao Received Researchers of DFH-1

In September 1969, all environmental simulation experiments for the prototype of DFH-1 satellite were completed, with all satellite systems meeting the requirements for the indicators and the technical status to be normal. In late October, Premier Zhou Enlai insisted on listening to the report on satellite development in details in person despite having much other work. When Qian Xuesen introduced Sun Jiadong to Zhou Enlai, Zhou Enlai shook hands with Sun Jiadong and said humorously: "Look how young the satellite expert is!" Upon hearing that, Sun Jiadong blushed and smiled, but was also put at ease.

The report consisted of Qian Xuesen's introduction to the development of "DFH-1" man-made satellite and the ongoing preparations at the time, as well as Sun Jiadong's explanation of the details in relation to the satellite.

Zhou Enlai listened to the report attentively and also asked the occasional questions. He was particularly concerned with the quality of each part of the satellite. After the report of Sun Jiadong, Zhou Enlai seemed to feel like interrogating the young Sun Jiadong when he asked the question: "How many cables are there on the satellite?"

▶ DFH-1 Satellite

Sun Jiadong answered the question according to the actual number of the cables on the satellite.

Then, Zhou Enlai continued: "How many plugs are there on the satellite?"

Sun Jiadong went red in the face and failed to answer the question. He could only say: "After I go back, I will make the statistics and report it to you, Mr. Premier."

Zhou Enlai said with smile: "You should take a careful and prudent attitude towards satellite development, and work like surgeons who are thoroughly familiar with each blood vessel and each acupuncture point of a patient before operating on that patient." Then, Zhou Enlai also emphasized: "Departments in charge of the satellite program should be modest to subordinate units instead of priding themselves on their seniority or oppressing others by power. Some departments used to have this problem, as they often stressed that they were working on 'national key projects' and overwhelmed others in the name of some central leaders. I hope this won't take place any more in the departments responsible for the satellite program. These departments should strengthen unity and help and coordinate with each other."

After that, Zhou Enlai thought for a while and then asked a serious question: "Once the satellite is launched, how will you demonstrate that it has really been sent into space? For example, upon the explosion of an atomic bomb, people can see the mushroom cloud. Since outer space is so large, how will you demonstrate to the world that China has succeeded in launching the first man-made satellite?" When Qian Xuesen had answered the question from several angles, Zhou Enlai nodded satisfactorily.

In March 1970, the assembly of the satellite and the carrier rocket was completed; on April 1st, the special train carrying the DFH-1 satellite and the LM-1 carrier rocket arrived at the Jiuquan Satellite Launch

Center. Those at the headquarters decided on the night of April 24th, 1970 for the satellite launch.

Early in the morning on April 24th, Mao Zedong gave approval for the satellite launch to be carried out. At 21:35 on the night, under order from the commander, bright flames spewed from the four engines of the first stage of the LM-1 rocket. Then, in a burst of loud rumbling, the rocket carrying the satellite pulled up slowly, soared into the sky and flew towards space. At 22:00, Zhou Enlai reported the good news to Mao Zedong. When the satellite flew past Kashgar 90 minutes after its launch, the radio at the Jiuquan Satellite Launch Center began to broadcast the music: "When the east is red, the sun rises…"

DFH-1, with a mass of 173 kg, operated in an orbit at a perigee altitude of 439 km and an apogee altitude of 2,384 km, with an inclination of 68.5 degrees and a cycle of 114 minutes. The satellite was a nearly spherical polyhedron with the diameter being about 1 m. It had silver zinc batteries as its power source and adopted aspin-stabilized approach with a spinning speed of 120 revolutions per minute (RPM). The top of the satellite was installed with a pair of ultra-short wave monopole antennae of 0.4 m long, and the satellite waist was equipped with 4 pairs of telescopic short-wave whip antennae of 3 m long.

With DFH-1 satellite's successful launch and operation in orbit, China became the fifth nation in the world to have succeeded in independently launching a satellite, following the Soviet Union, the United States, France and Japan. Based on DFH-1 and with more than four decades' development, China's space technologies have made tremendous progress. China's independently-developed aircraft and various communications, remote sensing, meteorological and navigation satellites have been applied in many fields and played vital roles in national defense, scientific research and the national economy. In addition, China has also sent astronauts into space via manned spacecraft

| Milestones in the Development of China's Space Industry |

and expanded its path to space through the launch of lunar exploration satellites.

▶ On-orbit Operation of DFH-1

▶ Internal Structure of DFH-1

▶ Joys over Successful Launch

▶ People throughout China Watching DFH-1 Fly into Night Sky

Exploration of China's Manned Spaceflight

When China's manned spacecraft sent its first astronaut into space, China's space cause took a huge firm step forward once again.

▶ Biological Experiments in Space: Early Exploration and Preparations for Manned Spaceflight

"The first person to dream of flying to the Moon was a beautiful Chinese girl and the first person to step onto the Moon was an American. That beautiful Chinese girl was Chang'e, while that American was me." In early summer of 1988, the American astronaut Neil Armstrong, the first person in the world to walk on the Moon, paid a visit to China. With these words, he began his speech at the Beijing Aerospace Medicine and Engineering Institute, which is currently known as the Astronaut Center of China (ACC).

In the face of unknown space, the independent initiative of humankind can never be replaced by machine. The exploration of mysteries in the universe, the development and utilization of space resources and each step of humankind toward space cannot be separated from direct human participation in the various activities in space. Spaceflights extend the scope of humankind's activities on the land, the ocean and in the atmospheric layer to space, and enable human beings to have a wider and deeper knowledge of the Earth and its surrounding environment, gain a better understanding of the universe as a whole, make full use of space and the special environment in manned spacecraft to conduct various experiments and research activities and exploit space and its rich resources. This is also the appeal of manned spaceflights.

However, when would Chinese astronauts go into space?

"We should first warm up the manned spaceflight program." Qian Xuesen said this when planning China's manned flight cause. In actual fact, China had secretly begun to make the preparations as early as the 1960s.

In 1958, CAS set up the Institute of Biophysics to undertake research on the effects of the spaceflight environment on human and living beings and on the protection methods as well. At that time, it was still a newly emerging field, and people had no real knowledge of it. Before carrying outmanned spaceflight, it was necessary to figure out some basic issues, for example, how much load can humans withstand during launch, liftoff, return and landing of a rocket; can humans adjust to weightlessness, and how will their body respond to the weightlessness; how can we protect astronauts from the danger of cosmic rays; and what effects will long-term spaceflights have on the human body? Answers to these questions require simulation experiments at ground level and, more importantly, biological experiments in space, which can provide

▶ Chairman Mao Having Dinner with Qian Xuesen

firsthand information.

On July 19th, 1964, China's first biological experiment satellite carrying an adult white rat entered into space, which was followed by two similar experiments with a flight altitude of 60-70 km on June 1st and 5th, 1965. In the latter two experiments, researchers placed in the head of each of the rockets 4 adult white rats and 4 young white rats, and 12 biological test tubes holding fruit flies, essential enzymes and other biological specimens. All the living organisms for these experiments were recovered safely, marking the complete success of the experiments.

Flight experiments with small animals were followed by similar experiments with larger animals. On July 15th, 1966, a dog named "Xiaobao" became the first large animal to travel to space onboard a T-7A (S2) biological experiment rocket and on July 28th, the dog Shanshan became the second large animal onboard a biological experiment rocket. During the flight experiments, ECG readings, blood pressure, breath, animal heat and other physiological parameters of the dogs were recorded, and their higher nervous activities were also observed with conditioned reflex experiment devices. The two dogs were in good health after they returned to Earth. These biological experiments onboard rockets had started China's biological experiments in space and constituted precious experience for the development of aerospace medicine.

In 1965, when studying China's first man-made satellite development plan, relevant departments began to discuss China's manned spacecraft program. From May 11th to 25th, 1966, CAS organized relevant military and civilian departments to convene a meeting to plan the development of a series of satellites over the years from 1966 to 1975. The meeting yielded the following planning idea: "With scientific experiment satellites as the starting point and with geodetic satellites, especially recoverable satellites, as the focus, first develop satellites for communications, weather, nuclear explosion detection, early warning and missile purposes, then form a comprehensive satellite application system and finally develop manned spacecraft based on recoverable satellites."

As to how to develop manned spacecraft, since recoverable satellites were emphasized as the key satellites for practical application, it was the logical choice to develop the first generation of manned spacecraft based on recoverable satellites.

▶ Sun Jiadong Talking with Qian Xuesen

▶ No Signs of "Dawn Light"

On January 8th, 1968, China held a meeting to show the general plan for China's first manned spacecraft. At the meeting, the first manned spacecraft was named Shuguang 1 (shuguang meaning "dawn light"). In April 1971, over 400 scholars, engineers and technicians from more than 80 entities nationwide gathered in Beijing to work on the demonstration spacecraft program, generally marking the start of China's manned spaceflight program, codenamed "Project 714". According to the spacecraft design proposal, Shuguang 1 spacecraft was derived from a recoverable satellite resembling a giant upside-down funnel in terms of its shape and comprised the cockpit and the equipment bay. The cockpit was equipped with two astronaut ejection seats, instruments and meters, radio communication

▶ Academician Wang Daheng

sets, control equipment and waste disposal units and furnished with drinking water, food, parachutes, etc. Shuguang 1 spacecraft was scheduled to be launched at the end of 1973.

In April 1968, the Aerospace Medicine and Engineering Institute was formed based on the Space Biology Department under the Institute of Biophysics, CAS. After that, China began the secrete selection of astronauts. In the preliminary screening, 88 pilots from the national air force were selected from more than 1,000 and asked to come to Beijing in the spring of 1971 for the final screening. The astronaut selection not only examined the physical condition of the candidates, but also tested their will and personal qualities. After strict tests for 12 training and test areas, including vibration, impact, centrifugation, low pressure and high temperature, China's first 19 astronaut candidates were finally determined.

However, things did not continue to go smoothly. For political, economic and technological reasons at that time, China suspended "Project 714" from October 1971. Since that point, this ray of "dawn light" was sealed for 20 years in the archive of CAST, until 1992.

▶ From the "863" Program to the "921" Program

By the 1980s, China's space technology had made rapid progress, as evidenced by its development of new models for the Long March carrier rocket series, its application of recoverable satellites and geosynchronous communications satellites, its possession of ocean space tracking ships and the formation of a relatively complete space engineering system, including the aircraft system, the space launch vehicle system, the space launch and recovery system and the satellite application system. In addition, the Aerospace Medicine and Engineering Institute founded in 1968 never ceased its aerospace medicine research.

In this period, there was no fundamental change to the situation in the Cold War, and major countries initiated strategy development and implementation plans one after another which centered on the development of high technology. Faced with this international situation, China's renowned scientists Wang Daheng, Wang Ganchang, Yang Jiachi and Chen Fangyun jointly presented to the central authorities "The Proposal on the Tracking and Study of the Development of Foreign Strategic High Technology" on March 3rd, 1986. On March 5th, Deng Xiaoping, Chairman of the Central Military Commission at that time, made the official comment that "This proposal is very important. Please discuss with experts and relevant responsible comrades, put forward detailed opinions and make a final decision. A quick decision should be made with regard to this issue, which cannot be delayed."

Soon after that, the State Council coordinated more than 200 experts and scholars from various fields to conduct specific research, and these experts and scholars worked out detailed plans and suggestions for

▶ Shenzhou Spacecraft Docking with Upper Stage of Long March Rocket

China's tracking of high and cutting-edge technologies in the world and submitted them to the Central Committee of the Communist Party of China (CCCPC) and the State Council on October 21, 1986. On November 18th, CCCPC and the State Council gave the formal approval to "The Outline for High Technology Development Plans". Since Deng Xiaoping made the official comment in March, the program was coded "863".

When approving the 863 program, the central authorities also decided to allocate a special fund of RMB 10 billion for the implementation of the strategy, which was a figure far beyond the expectations of the scientists. Academician Yang Jiachi said when recalling the state's investments in the 863 program: "On one morning, Mr. Zhang Jingfu called the four of us to meet at Zhongnanhai and talk over the problem. He asked us how much of a research and development fund we would need. Since Wang Ganchang has some of experience in these matters, he said we needed RMB 10 million. Upon hearing that, Zhang Jingfu responded immediately and said that amount was absolutely not enough and the state approved RMB 10 billion. The four of us were completely taken aback by that figure." In the 863 program, pre-research on manned spaceflight enjoyed the greatest priority as a key development program, and 4 billion of the 10 billion was designated to be used for space technology.

Formation of the 863 program acted as a catalyst in China's commencement of manned spaceflight exploration. After decades of hard development, China's manned spaceflight cause already stood at a new starting line.

On September 21st, 1992, China decided on the Three-Step Development Strategy for the manned spaceflight program. The first step was the manned spacecraft stage, and in this stage, China should launch unmanned and manned spacecraft, send astronauts into Low-Earth Orbit (LEO) and enable them to return safely and conduct earth observation and space application experiments; in the second step, namely, the space experiment stage, China should make further breakthroughs in manned spaceflight technologies, including technologies for multi-manned and multi-day spaceflights, astronauts' landing and walking in space, rendezvous and docking between manned spacecraft and space labs, cargo spaceships ferrying supplies to space labs and prolonged sustaining of life in space labs, constructing space labs and solving issues concerning the application of space labs of a certain size and with short-term manning; in the third step, that is, the space station stage, China will build manned space stations and solve problems in relation to the application of larger-scale space stations with long-term manning. It was on this day that China's manned spaceflight program was officially established and codenamed 921.

Currently, China's manned spaceflight program has entered into the second stage, as it has launched ten Shenzhou spacecrafts and Tiangong 1 (tiangong literally meaning "heavenly palace") target module and carried out astronaut spacewalks and space rendezvous and docking missions.

▶ China's Manned Spaceflight: A Huge and Systematic Program

China's manned spaceflight program has been the largest and most technologically difficult program with some of the most complicated systems in the history of its space development. The program consists of the general engineering plus seven major systems, which are the astronaut system, the space application system, the manned aircraft system, the carrier rocket system, the launching system, the manned spaceflight monitoring, control and communications system and the landing system.

The astronaut system deals with selection and training of astronauts, provision of medical checks and safeguards for them in outer space, research and development of space suits and relevant equipment able to perform medical checks on astronauts in the process of spaceflight which can then transmit this data and inclusion of medical specifications in the engineering design of the spacecraft. In addition, this system also encompasses in-flight environmental control for the astronauts, and the environmental control and life-support subsystem needed to create an atmospheric environment for astronauts suitable for both life and work.

The space application system carries out space science and application research. The scientific experiment instruments housed in the spacecraft can conduct space-to-ground surveillance and various scientific experiments, with the research findings able to be widely used in pharmaceutical development, food and health, prevention and control of difficult and complicated diseases, agriculture, etc.

The manned spacecraft system undertakes the development of Shenzhou manned spacecraft. After the spacecraft orbits as planned and the spaceflight missions are fulfilled, the re-entry capsule will return to Earth along the predetermined route and the orbital module will remain in space for half a year to carry out space-to-ground surveillance and other scheduled experiment tasks.

The carrier rocket system assumes development of the LM-2 carrier rocket that sends the spacecraft into space. LM-2F is currently the most reliable and secure carrier rocket in China. It boasts a 99.9% launching reliability. The carrier rocket must deal with range launch, transportation, error detection and life-saving astronaut escape issues.

The launching system deals with test and launch of the spacecraft and the rocket and surveillance and control tasks in the liftoff stage, which is in the charge of the manned spaceflight launching ground at the Jiuquan Satellite Launch Center. The launching ground constitutes a technical

▶ "Shenjian" and "Shenzhou" Ready for Launch

▶ Reentry Module and Parachute of Shenzhou Spacecraft

China's Journey to Space: From Dream to Reality

zone, a launch zone, a test control zone, etc. The launching ground that entered into service in 1998 adopts the internationally advanced "three vertical" technologies, namely, vertical assembly, vertical test and vertical transportation, and the long distance test and launch mode.

The manned spaceflight monitoring, control and communications system monitors and remotely controls the launch, orbit and return of the spacecraft. It is also the only communication link between the spacecraft and ground control after the spacecraft has been sent into space. The system is currently equipped with six Yuanwang (yuanwang literally

▶ On-orbit Flight of Shenzhou 5 Spacecraft

▶ Reentry Module of Shenzhou 5 Spacecraft

meaning "long vision") space tracking ships, six measuring stations on land and three mobile measuring stations.

The landing system involves the capture, tracking and measurement of the spacecraft upon its re-entry, the search and recovery of the re-entry capsule and medical checks, care and aid to astronauts after their return back to Earth. It includes the master landing ground in the middle of Inner Mongolia and the backup landing ground at the Jiuquan Satellite Launch Center, as well as several emergency rescue zones on standby including land and sea areas.

A New Journey for Lunar Exploration

"China has fulfilled its dream of flight to the Moon, while Chang'e also seems to be celebrating the great moment by sprinkling sweet scented osmanthus wine upon the Earth." When China's first lunar exploration satellite succeeded in entering orbit around the Moon, China's space undertaking began to step into deeper and wider space.

▶ "Seed" of the Dream

On May 28th, 1978, Zbigniew Brzezinski, the United States National Security Advisor, visited China and was received by President Hua Guofeng. As part of his visit and as commissioned by the US President Jimmy Carter, Brzezinski presented China with a precious gift: a small piece of lunar rock sample collected by American astronauts during the United States' Apollo lunar landing. The stone, as small as a fingertip and only weighing 1 gram, was cast in an organic glass box resembling

▶ Insignia of China Lunar Exploration Program

a convex lens.

After receiving the gift, President Hua Guofeng handed it to the scientific research department and required it to conduct relevant research on the stone. At that time, there were few researchers studying celestial rock and only the Institute of Geochemistry, CAS (Guiyang) was doing relevant research. Therefore, the small stone was sent to Guiyang, where Ouyang Ziyuan, still a researcher at that time, led the research. Although the Americans revealed no information regarding this rock, it was just like a "seed" of dreams, which took root and spread in the heart of the Chinese scientist Ouyang Ziyuan. Now, more than thirty years has passed and Ouyang Ziyuan, who is now an academician with CAS, still keeps the Moon as the subject of his scientific investigations.

During initial research and tracking of foreign lunar exploration activities, a group of scientists, which included Ouyang Ziyuan, also thought to suggest that relevant national departments begin developing and launching lunar probes. However, in view of the current status of the state of affairs in the country, its economic condition and the space technology level at that time, China had not yet developed the capacity to carry out lunar exploration.

When would the dream of lunar exploration come true? That was the call from the bottom of Ouyang Ziyuan's heart.

▶ **Ten Years' Demonstration**

Completion of a major scientific project requires not only the spirit of persistent pursuit, but also long-term planning and clear objectives. Song

Jian, Vice Chairman of the CPPCC National Committee and President of CAE at that time, had written to Ouyang Ziyuan, encouraging scientists to work on China's lunar exploration development planning, program design and scientific research and advising them to have clear scientific and engineering objectives.

In the 1990s, major space countries represented by the United States started a new round of lunar exploration, creating a great stir among Chinese scientists. "When competition in lunar exploration around the world becomes white hot, if China does not take action, it will lag behind the competition and lose the right to speak." Ouyang Ziyuan said.

In 1994, experts led by Ouyang Ziyuan completed China's first feasibility report for lunar exploration, discussing the necessity and feasibility of China conducting lunar exploration and commencing ten years' of debate regarding China's lunar exploration program.

In 1995, the aerospace expert committee of the 863 program proposed the research subject of "The Necessity and Feasibility of China's Conducting Lunar Exploration". For this research subject, Ouyang Ziyuan, Ye Zili, Chen Kangwen, Chu Guibai, Lin Wenzhu and other scientists cooperated in developing the first relatively complete set of lunar exploration demonstration reports, including "The Necessity and Feasibility of China Conducting Lunar Exploration", "The Study on the Feasibility of the Lunar Satellite Technology Scheme", "The Study on Key Technologies in the Lunar Satellite Program", etc.

In 1998, when China carried out major reforms of government institutions, the State Council established the new Commission of Science, Technology and Industry for National Defense (COSTIND) under which there was the China National Space Administration (CNSA). The establishment of COSTIND played a role promoting the speeding-up of the debate over the lunar exploration program.

In August 2000, the expert conference organized by CAS approved the research entitled "Scientific Objectives and Payload of the Lunar Exploration Satellite", marking the formal establishment of scientific objectives of the first-stage project of China's lunar exploration program. In more certain terms, scientific objectives of the first-stage project included drawing three-dimensional lunar surface images, analysing the contents of 14 elements on the lunar surface and their distributions, exploring the features and thickness of the lunar soil and exploiting the space environment between the Earth and the Moon.

In November 2000, the State Council Information Office published the white paper entitled "China's Space Activities". It was clearly pointed out in the paper that China would "conduct pre-research on deep space exploration centering on lunar exploration". This attracted so much attention from Chinese and foreign media that for a long time after that, "China's lunar exploration" was one of the most commonly used phases across all media.

In December 2001, COSTIND entrusted academician Sun Jiadong with the coordination and development of the framework for China's lunar exploration program. Sun then convened lunar scientists and key figures in the field of aerospace from all over China so they could devote themselves to completely expound and verify the feasibility of the first-stage project of the lunar exploration program, which lasted more than a year.

At the end of 2003, driven by the academician Song Jian, CAE convened an academician symposium on China's lunar exploration program. All of the more than 20 academicians present at the symposium agreed that China should have its place in lunar exploration and deep space exploration and that China's lunar exploration program would not only promote the development of China's basic subjects of science and the progress of China's aerospace technology, but also play a stimulating role in the development of national economy.

▶ Realization of the "Dream of Lunar Exploration"

In January 2004, the first-stage project of China's lunar exploration program, the lunar orbiter project, was officially launched, symbolizing that China had begun to realize its dream of lunar exploration.

For the project, COSTIND established a lunar orbiter project leadership group and officially named the lunar exploration program the "Chang'e program" and the first lunar exploration probe "Chang'e 1". After that, COSTIND also set up two lunar orbiter project and scientific research divisions with chief commander Luan Enjie, chief designer Sun Jiadong and chief scientist Ouyang Ziyuan as core members, formed the lunar exploration program center and built an engineering system which involved nationwide collaboration.

On the basis of an analysis of the achievements of foreign countries in lunar exploration and future lunar exploration programs of various countries around the world and with a view to China's own scientific and technological level at that time, its own internal state of affairs and its general growth strategies, China decided to concentrate on unmanned lunar exploration before 2020. It would achieve this in three steps: lunar orbiters, lunar soft landers and lunar rovers and lunar sample returns, which can be summarized as "orbiting, landing and returning".

Currently, China has completed the first-stage project, the lunar orbiter project, made breakthroughs in Earth-to-Moon flight, long distance surveillance and control, lunar orbiter, remote lunar monitoring and analysis, and carried out the four major scientific objectives in this stage: acquisition of three-dimensional lunar images, analysis of the contents of lunar elements, substance types and their distribution features, exploration of lunar soil features and thickness and exploitation of the space environment between the Earth and the Moon.

Now, China is working on the second-stage project of the program, that is, the soft lander project. It will launch a soft lander carrying a lunar rover for soft landing on the Moon and on-site exploration around the landing zone. The second-stage project will achieve five major scientific objectives: investigation and analysis of the landscape and geological formation of the landing zone, inspection of substances and available resources on the Moon's surface, probe into the internal structure of the Moon, monitoring the space environment between the Sun, the Earth and the Moon and Moon-based astronomical observations.

The third-stage project is lunar sample return. In this stage, China will send a small lunar sampling and returning capsule, a lunar rock borer and a lunar robot for collecting key samples and returning them to Earth and for performing fine study on the landing zone in order to provide reliable data for manned lunar exploration and lunar base construction in the future.

China's lunar exploration program is a complicated and systematic one with six major systems: the lunar probe system, the carrier rocket system, the monitoring, control and communications system, the launching system, the ground application system and the landing and recovery system.

The lunar probe system sends various scientific instruments within proximity of the Moon or enables them to land onto the Moon, provides scientific exploration instruments with necessary flight attitude assurance, installation position, vision, energy, temperature environment, data management and other conditions, and returns the exploration data acquired to Earth or brings lunar samples collected on the surface of the Moon back to Earth. It includes lunar probes (lunar exploration satellites), lunar soft landers, lunar rovers (i.e. mobile lunar vehicle) and lunar sampling and returning capsules. China succeeded in launching Chang'e 1 and Chang'e 2 lunar exploration satellites in 2007 and 2010 respectively.

The carrier rocket system is a launch vehicle to send a lunar probe away from the Earth to a certain orbit altitude and to help it gain a certain speed. Since it has to provide the lunar probe with a flight speed close to the second cosmic velocity, it needs to have a larger thrust and a greater carrying capacity. To date, China has launched two lunar exploration satellites on an LM-3A carrier rocket.

The monitoring, control and communications system acts as an information bridge between the lunar probe and the Earth and it is for carrying out tracking, remote monitoring, remote control and data transmission of the carrier rocket and the lunar probe, understanding flight status of the lunar probe, issuing commands to the lunar probe and completing scheduled flight missions.

The launching system provides the site for docking, test and launch of the lunar probe and the carrier rocket.

The ground application system deals with practical application of information acquired by the lunar probe and applied research. It converts exploration data into scientific discoveries and thus promotes the development of space science.

The landing and recovery system is a new system added to the third stage of the lunar exploration program. It is responsible for searching the landing site of the re-entry capsule of the automatic lunar sampling and returning probe and taking lunar samples out of the re-entry capsule and sending them to labs as soon as possible.

Through implementation of the lunar exploration program, China will make key technological breakthroughs in unmanned lunar exploration, achieve a series of independent and innovative results and gain technological strength in deep space exploration while laying a solid foundation for sustainable development of space science research and deep space exploration.

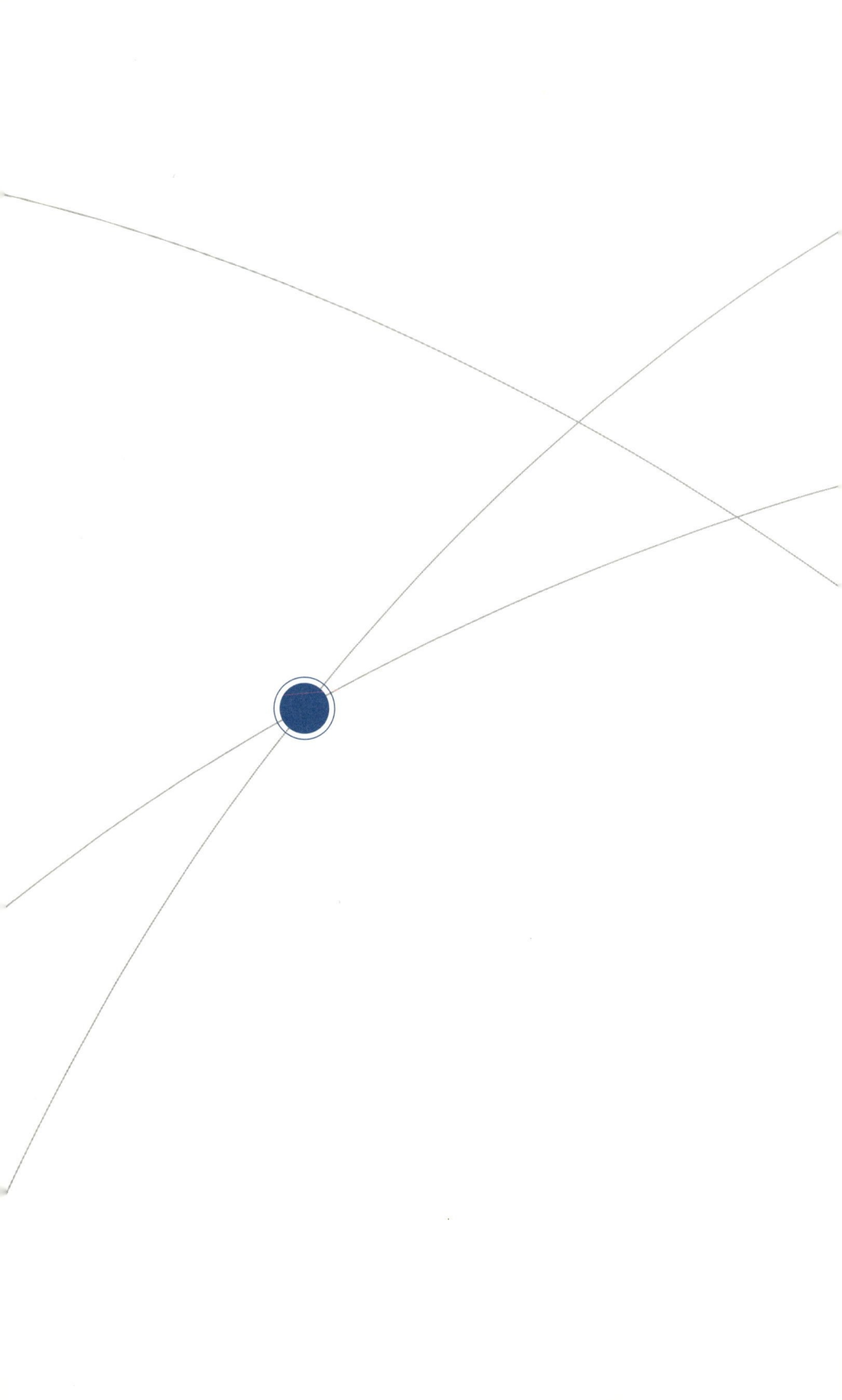

2

Long March Rocket Family

China has always been hailed as the home of rockets. As early as the 11th century A.D., people in the Song Dynasty invented the first gunpowder-propelled rockets in the world, which spread to Arab and western countries in the 13th century.

It was not until the mid-1960s that China began to develop its modern launch vehicles. After tough research efforts made by Chinese scientists, China launched its Dongfanghong 1 (DFH-1) satellite into the near-Earth orbit on its Long March 1 (Changzheng 1) carrier rocket on April 24th, 1970, becoming the fifth nation in the world to have successfully launched a domestic satellite on its independently developed carrier rocket. Over the past 40-odd years, China has kept to the road of "self-dependence and independent innovation" for the development of carrier rocket technology and has accomplished tremendous achievements, greatly promoting the development of China's satellite and satellite application technology, manned space technology and lunar exploration technology.

China has successfully developed Long March 1 (LM-1) series, Long March 2 (LM-2) series, Long March 3 (LM-3) series and Long March 4 (LM-4) series of carrier rockets, which form a Long March rocket family made up of 14 models. These rockets equip China with the capability to launch different kinds of satellites, manned spacecraft and deep space probes into low, medium and high Earth orbits, with a carrying capacity of 12 tons for LEO, a sun synchronous orbit carrying capacity of 6 tons and a geostationary transfer orbit carrying capacity of 5.5 tons. In addition, China has mastered the technologies for the separation of multi-stage rockets, multi-satellite launch with a single rocket, separation of large rockets with strap-on boosters, large spacecraft fairing, low-temperature and high-power rocket engines and high-altitude secondary rocket ignition. Moreover, China has seen the accuracy of its orbit injection rockets reach internationally advanced levels and can meet the demands of various international satellite users.

To date, China's Long March rocket family has conducted more than 100 launches, with a rocket launch success rate of 93%.

LM-1 Carrier Rocket Series: China's First Carrier Rocket Models

The LM-1 is the first member of China's Long March family, and it was mainly used for launching small LEO satellites. China began development of LM-1 in 1965, which took more than four years and covered the four major systems: power, structure, control and flight measurement. The rocket as a whole and its subsystems adopted many new technologies, some of which approached and even reached internationally advanced levels in the 1960s. The LM-1's first and second stages adopted liquid propellant rocket engines adapted from those of the intermediate and long-range missile, and the third stage used a solid propellant rocket engine.

LM-1 had a total length of 29.8 m, a maximum diameter of 2.25 m, a liftoff mass of 81.5 tons, a liftoff thrust of 1,020 kN and a carrying capacity for sending a payload of 300 kg to a circular orbit with an inclination of 70 degrees and at an altitude of 440 km.

On April 24th, 1970, LM-1 placed the first DFH-1 satellite into orbit, initiating a new

▶ LM-1 Rocket to Launch DFH-1 Satellite

era in the development of China's space industry. On March 3rd, 1971, LM-1 sent the Practice 1 (Shijian 1) scientific experiment satellite into pre-selected orbit, which was the second launch and also final flight of the rocket.

LM-2 Carrier Rocket Series: Sending Shenzhou Spacecraft into Space

China started the development of the LM-2 carrier rocket series in 1970. Mainly used for launching low-Earth orbiters, LM-2 consists of LM-2 (first launched in 1974), LM-2C and its improved version (first launched in 1982), LM-2D (first launched in 1992), LM-2E (first launched in 1990) and LM-2F (first launched in 1999).

▶ LM-2

LM-2 is the base model of the LM-2 rocket series and a two-stage launch vehicle derived from a Chinese intercontinental missile. With a total length of 31 m, a maximum diameter of 3.3 m, a liftoff mass of 192 tons and a liftoff thrust of 2,700 kN, LM-2 was able to place a payload of up to 1,800 kg into LEO at an altitude of 200 km. This rocket was equipped with the advanced gimbaled rocket engines, a higher-power hydraulic servo system and a newly-developed platform computer that greatly enhanced the guidance precision, reliability and carrying capacity of the rocket. On November 26th, 1975, LM-2 accurately sent China's first recoverable satellite into pre-selected orbit, followed by the successful launch of China's second and third recoverable satellites. In 1979, LM-2 was retired.

▶ LM-2C

In the late 1980s, China's carrier rockets began to stride forward towards the international commercial launch market. From 1987 to 1988, LM-2C sent microgravity test instruments from France and Germany into space on China's recoverable satellite.

In April 1993, China and the US inked the contract for launching American iridium satellites. For this reason, China began to develop the improved version of LM-2C for launch of iridium communications satellites. From 1997 to 1999, the improved model conducted 7 consecutive launches, successfully sending 2 analog satellites and 12 iridium satellites and satisfactorily accomplishing the tasks specified in the contract.

▶ LM-2C Rocket

The improved version of LM-2C was a three-stage carrier rocket, with a total length of 43 m, a maximum diameter of 3.3 m, a liftoff mass of 245 tons, a liftoff thrust of 2,900 kN and a capability of sending a payload of 1.9 tons into sun synchronous orbit at an altitude of 600 km.

▶ LM-2E

In view of the increasing demands on the international commercial launch market as a result of the market's rapid development, China also

▶ LM-2E Rocket

developed LM-2E, which is the first strap-on rocket developed by China and is also known as "LM-2E strap-on rocket".

LM-2E has a total length of 49 m, a maximum diameter of 11 m, a liftoff mass of 462 tons, a liftoff thrust of 5,923 kN and a launch capability of sending a payload of 9.2 tons into LEO at an altitude of 200 km. This rocket could undertake large-scale communications satellite launch missions. In 1987, the demonstration program for LM-2E was started. It took four years, but the actual development only lasted 18 months. On July 16th, 1990, LM-2E successfully carried out its maiden test flight. Since 1992, LM-2E has made 7 launches in total and undergone severe challenges.

In 1995, LM-2E exploded during the launch of the USA-made Apstar 2 communications satellite, leading to the destruction of both the rocket and the satellite. After a series of rectifications, LM-2E stepped out of the shadow of failure and launched the USA-made AsiaSat 2 and EchoStar 1 communications satellites on November 28th and December 28th, 1995 respectively.

▶ LM-2F

From November 20th, 1999 when LM-2F launched the Shenzhou 1 spacecraft to June 11th, 2013 when LM-2F sent the Shenzhou 10 spacecraft into space, LM-2F maintained an excellent performance record of successfully achieving each of the ten launch missions, making huge contributions to the smooth implementation of China's manned spaceflight program. It is for this reason that LM-2F is also known as "Shenjian" (shenjian meaning "divine arrow").

LM-2F is a highly reliable carrier rocket with a large thrust developed from LM-2E according to requirements of China's manned spaceflight missions. This new large strap-on and two-stage liquid fuelled carrier rocket is the rocket with the largest liftoff mass, the longest length and the most complicated systems among all of the current carrier rockets of China, and it is mainly used in launches of Shenzhou spacecraft, and large-scale target aircraft, into LEO.

In 1992, China's manned spaceflight project was formally launched, which initiated the eight-year long development of LM-2F accordingly. LM-2F, as the first manned rocket model of China, put forward specific reliability and safety design objectives for the first time, inherited its main configuration from LM-2E, adopted more than 50 new technologies and involved over 100 key technological breakthroughs, which ensure the safe and reliable separation of the rocket from spacecraft and bring its stability and reliability to internationally advanced levels.

With a total length of 58.3 m, a maximum diameter of 10 m (including boosters), a liftoff mass of 479 tons and a liftoff thrust of 5,923 kN, LM-2F can place a payload of up to 8 tons to LEO at an altitude of 200 km and to high Earth orbit (HEO) at an altitude of 350 km. The rocket consists of 10 subsystems, including launcher structure, power equipment, control system, propellant utilization system, fault monitoring management system, launch escape system, remote monitoring system, external safety monitoring system, ground facilities and auxiliary system, with the fault monitoring management system and the launch escape system being unique to the manned rocket.

The greatest difficulty faced by the fault monitoring management system is to avoid error or unnecessary flight abortion. If there is nothing wrong with the rocket itself, and the fault monitoring management system makes an incorrect assessment and issues the abort command, the astronauts may be safe, but the whole flight will have failed. If there is something wrong with the rocket itself, but the fault monitoring management system fails to find the problem and issue the abort command, it would lead to the rocket and spacecraft crashing completely and, more severely, it would cost the precious lives of astronauts.

▶ LM-2F Rocket

The abort engine of this rocket was the first one independently developed by China and it meets the quality objective of a 100% pass rate for each delivery, after achievement of numerous technological breakthroughs, development of more than 100 engine specimens of four configurations and completion of ground hot firing tests for a dozen of the engines. In 1998, China finally completed the ground-level abort flight test for the launch abortion system, demonstrating that China's independently developed solid-fuelled engine for the launch abortion system had reached internationally advanced levels in terms of its comprehensive performance.

LM-3 Carrier Rocket Series: Ladder to the Moon

In the late 1970s, China began to develop the LM-3 carrier rocket series, which includes the four models of LM-3, LM-3A, LM-3B and LM-3C and primarily undertakes launches of geosynchronous high-altitude orbit aircraft.

▶ LM-3

LM-3 has a total length of 44 m, a diameter of 3.35 m for the first and the second stages and of 2.25 m for the third stage, a liftoff mass of 202 tons, a liftoff thrust of 2,962 kN and the capability of sending a payload of 1.6 tons into geosynchronous transfer orbit (GTO).

China started schematic design of LM-3 in 1978, entered the preliminary design phase in 1980 and performed the first launch on January 29th, 1984 after six years' development. On April 8th,

▶ Long March Rocket Family

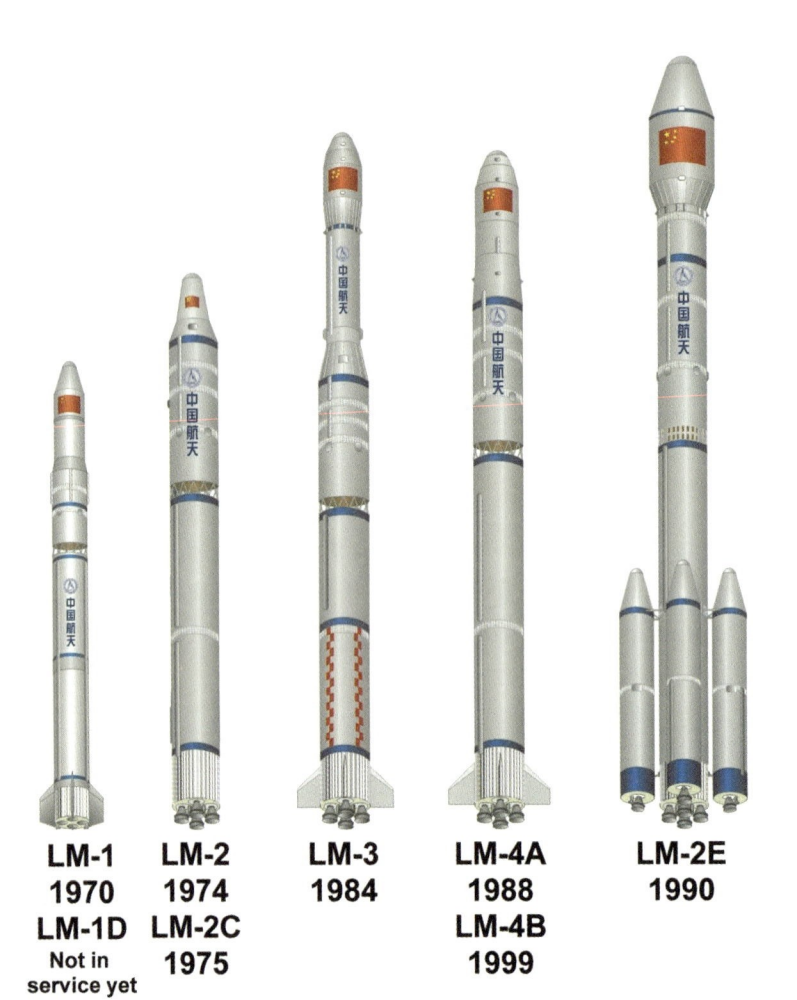

LM-1
1970
LM-1D
Not in service yet

LM-2
1974
LM-2C
1975

LM-3
1984

LM-4A
1988
LM-4B
1999

LM-2E
1990

LM-2F
1992

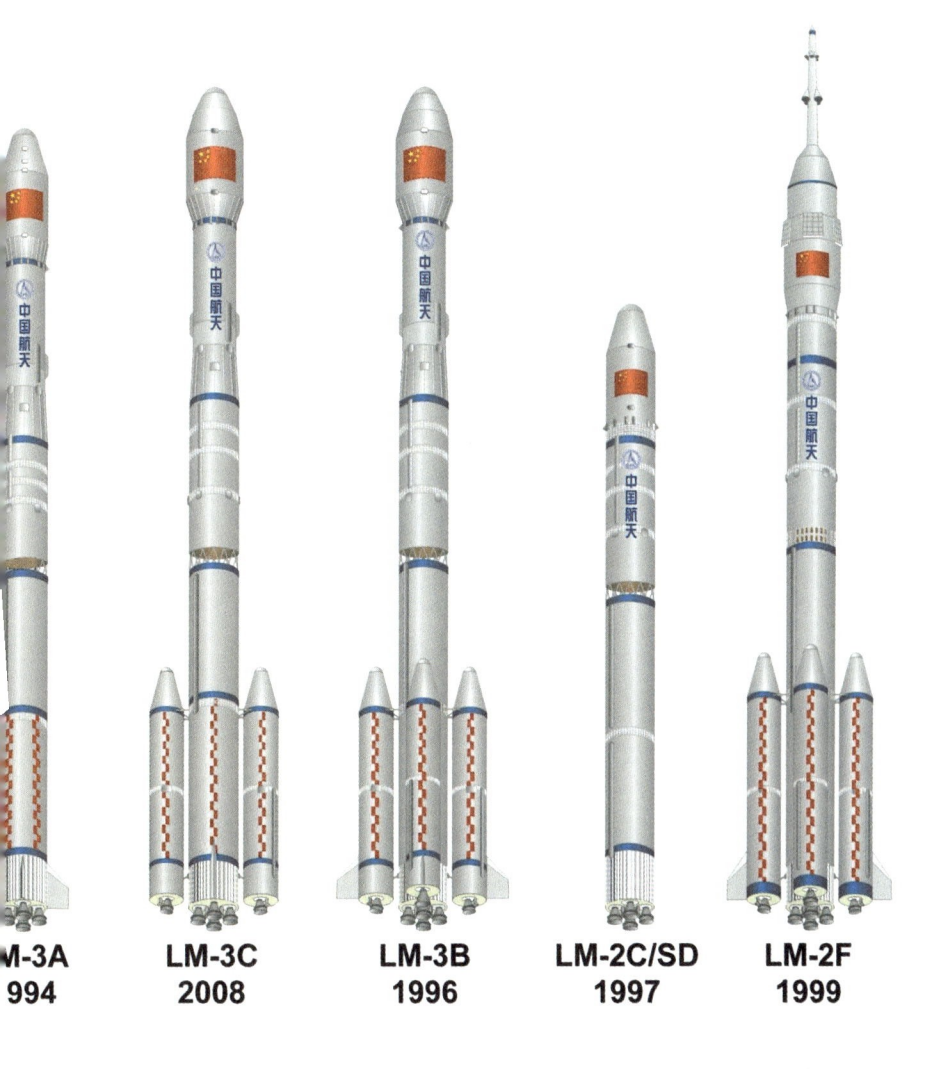

LM-3A	LM-3C	LM-3B	LM-2C/SD	LM-2F
1994	2008	1996	1997	1999

1984, LM-3 succeeded in sending China's first geostationary test communications satellite DFH-2 into space, showing that China's carrier rocket technology had reached new heights.

From February 1986 to June 2000, LM-3 conducted 11 launches. In particular, on April 7th, 1990, LM-3 successfully sent the AsiaSat 1 manufactured by Hughes Aircraft Company into predetermined orbit during its seventh launch. This was the first time China had succeeded in launching a foreign communications satellite and it marked the formal entry of Chinese carrier rockets into the international commercial launch market. LM-3 enjoyed the advantage of low launch costs in the international market at that time, with the launch cost of each rocket being about USD 35 million (the price in 1993 and 1994). Therefore, this launch of a foreign satellite had great international influence. The *Christian Science Monitor* pointed out in its comment on April 9th of the year that "The launch of the AsiaSat 1 satellite represents a significant diplomatic as well as technical achievement for China."

▶ **LM-3A**

In October 2007, LM-3A launched Chang'e 1, China's first independently developed lunar exploration satellite into space, enabling it to embark on its journey to the Moon. LM-3A is a geosynchronous, three-stage carrier rocket with a greater carrying capacity and was newly-developed after the successful launch of LM-3. The demonstration program for LM-3A began in March 1985 and its formal development

▶ LM-3A Rocket

| Long March Rocket Family |

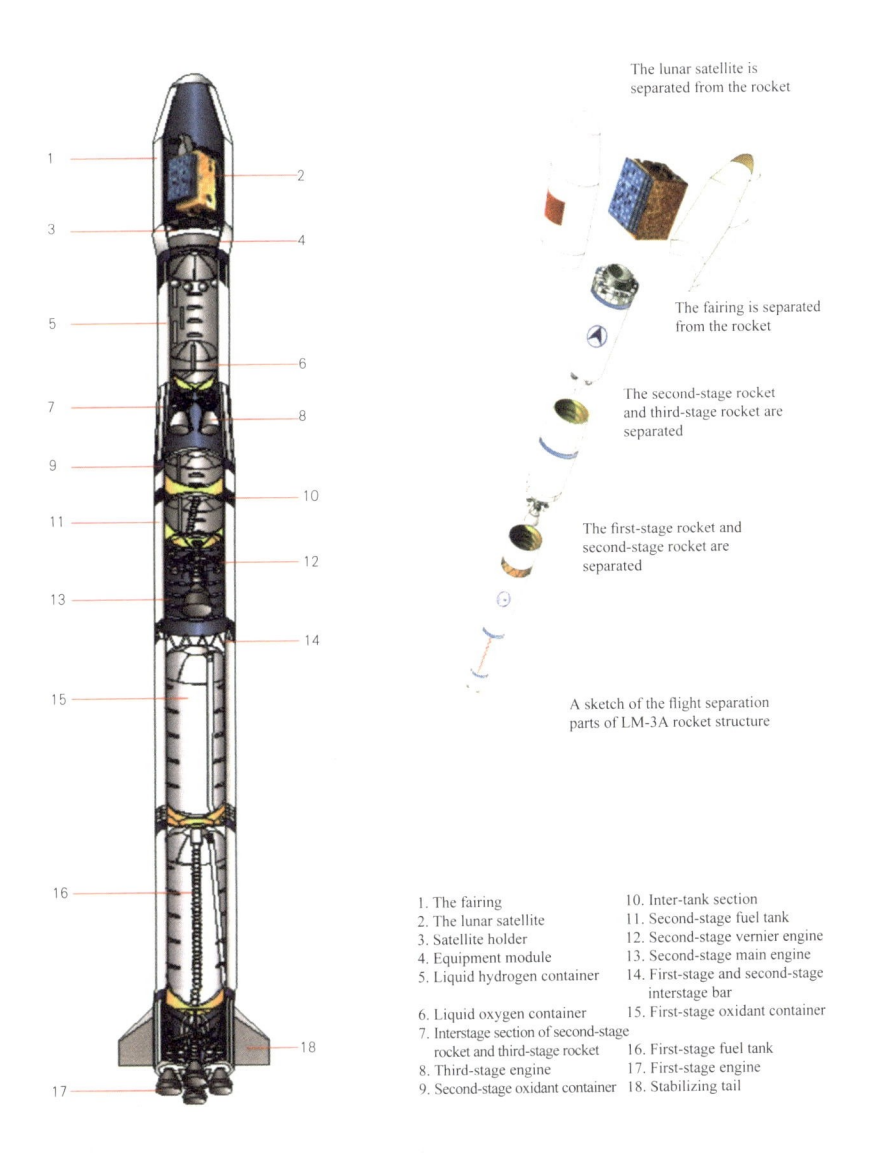

The lunar satellite is separated from the rocket

The fairing is separated from the rocket

The second-stage rocket and third-stage rocket are separated

The first-stage rocket and second-stage rocket are separated

A sketch of the flight separation parts of LM-3A rocket structure

1. The fairing
2. The lunar satellite
3. Satellite holder
4. Equipment module
5. Liquid hydrogen container
6. Liquid oxygen container
7. Interstage section of second-stage rocket and third-stage rocket
8. Third-stage engine
9. Second-stage oxidant container
10. Inter-tank section
11. Second-stage fuel tank
12. Second-stage vernier engine
13. Second-stage main engine
14. First-stage and second-stage interstage bar
15. First-stage oxidant container
16. First-stage fuel tank
17. First-stage engine
18. Stabilizing tail

▶ Structural Diagram of LM-3A Rocket

started in April 1986. Compared with LM-3, LM-3A has significantly improved technological and service performance and adaptability.

With a large enough carrying capacity and multiple ignition capability, LM-3A is able to launch lunar exploration satellites. With a total length of 52 m, a maximum diameter of 3.35 m, a liftoff mass of 241 tons and a liftoff thrust of 2,961 kN, LM-3A can haul a payload of 2.6 tons into GTO. Since the first launch conducted on February 8th, 1994 at the Xichang Satellite Launch Center, LM-3A has maintained a launch success rate of 100%, for which it has the honor of being the "Gold Medal Rocket" of China.

▶ LM-3B Rocket

▶ LM-3B

LM-3B had its overall plan design initiated in July 1989 and it made its maiden flight in 1996. With a total length of 56 m, a maximum diameter of 3.35 m, a liftoff mass of 456 tons and a liftoff thrust of 5,923 kN, the rocket can place a payload up to 5.5 tons into GTO, conduct multi-satellite launches with a single rocket or launch satellites into other orbit. To date, LM-3B has successfully launched many large-scale satellites, including Agila-2 of Philippines, Apstar 2R, ChinaStar 1 (Zhongwei 1), SinoSat 1 (Xinnuo 1) and Nigeria's NigcomSat 1R satellite, etc.

▶ LM-3C

On October 1st, 2010, the LM-3C carrier rocket successfully sent Chang'e 2 lunar exploration satellite into per-determined orbit at the Xichang Satellite Launch Center, becoming another "ladder to the Moon". Derived from the established LM-3A and LM-3B, LM-3C also made bold innovations as the first asymmetric rocket in China. Compared with other carrier rockets, although LM-3C has not undergone many practice tests, it is an improved version of the previous established rockets and features advanced technological performance, high reliability, moderate carrying capacity and strong adaptability, making it the first choice for the Chang'e 2 lunar exploration mission.

With a total length of 56 m, a maximum diameter of 8 m, a liftoff mass of about 367 tons and a liftoff thrust of 4,442 kN, LM-3C comprises the launcher structure, power system, control system, measurement system, glide propellant management and attitude control engine system, propellant utilization system and auxiliary system. Among them, the control system adopts all digital attitude control design, which ensures strong adaptability and high reliability of the system and

China's Journey to Space: From Dream to Reality

the measurement system features a high integration degree and excellent performance.

In January 2003, China's first data relay satellite program was established, which was officially named Tianlian 1 (tianlian meaning "Sky Link") and chose LM-3C as its matching rocket. To meet demands of diverse launch markets, LM-3C is equipped with two boosters, which was the first rocket to have such a configuration in China.

LM-3C is an important member of the LM-3 series, and its successful development marks LM-3 forming a complete series as well as a rocket group with the largest high orbit carrying capacity and strongest adaptability in China.

▶ LM-3C Rocket

LM-4 Carrier Rocket Series: Sending Satellites into Sun Synchronous Orbit

The LM-4 carrier rocket series includes LM-4A, LM-4B and LM-4C, and primarily undertakes the launch of satellites into sunsynchronous orbit.

On September 7th, 1988, the first LM-4A carrier rocket succeeded in placing China's first experimental weather satellite into sunsynchronous orbit. LM-4A has a total length of 41 m, a maximum diameter of 3.35 m, a liftoff mass of 241 tons, a liftoff thrust of 2,961 kN and the ability to send a payload of 1.5 tons into sunsynchronous orbit at an altitude of 901 km.

The LM-4B is an improved model of the LM-4A with a large carrying capacity and strong adaptability, and its schematic design began in 1989. The rocket can be installed with a satellite fairing with a diameter of 2.9 m, 3.35 m or 3.8 m, making it able to meet the requirements of different satellites and perform various orbital launch missions. In addition to single satellite launch, it can also undertake multi-satellite launches and lift-launch of microsatellites. With its total length extended to 48 m and its liftoff mass increased to 249 tons, the rocket can place a payload of 2.6 tons into orbit. On May 10th, 1999, LM-4B successfully launched a Fengyun 1 (FY-1, fengyun meaning "wind and cloud") weather satellite and a Practice 5 scientific experiment satellite into space.

With schematic design commencing in 1999, the LM-4C made its maiden flight in April 2006. The rocket continues to implement the design ideas of commonality, serialization and modularization. The third-stage engine adopts a two-start operation mode, doubling the carrying capacity. At the same time, with enhancement of the capability of multi-satellite launches with a single rocket, it can meet the launch requirements of a variety of satellites.

China's Journey to Space: From Dream to Reality

▶ LM-4C Rocket

| Long March Rocket Family |

▶ Long March Series of Rocket Models Exhibited at Zhuhai Air Show

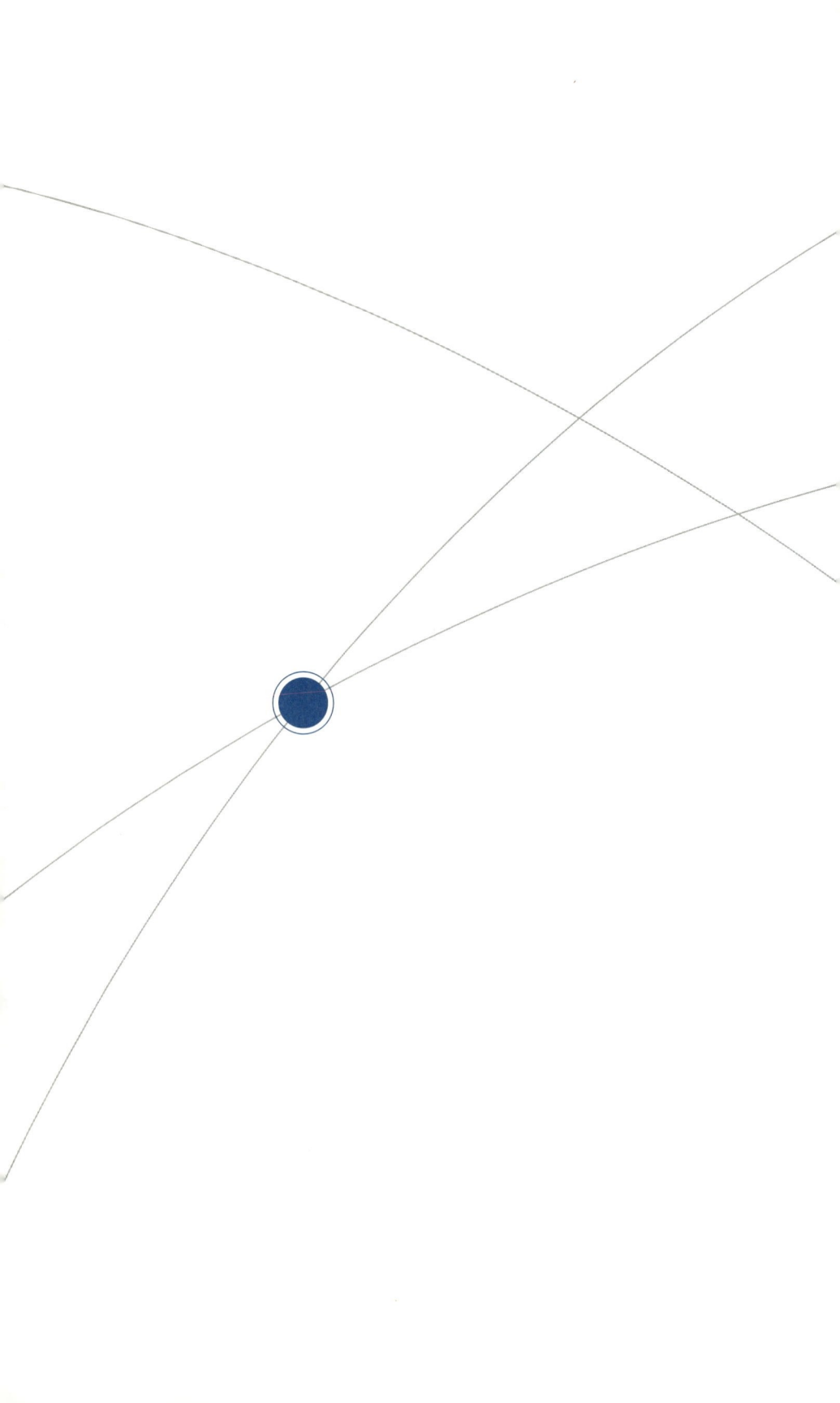

3

Chinese Satellites in Space

In the more than four decades since the successful launch of China's first satellite DFH-1 in 1970, China has made tremendous progress in its satellite development technology. By the end of 2012, China had successfully developed and launched 168 man-made satellites (excluding exported satellites), gradually formed a complete satellite series covering earth observation, communications and broadcast, navigation and positioning and science and technology experiment satellites, and fulfilled the transformation from applied satellites to service satellites. These satellites have been widely used in social, economic, scientific and technological, cultural, educational and other fields.

Widely-used Earth Observation Satellites

▶ Recoverable Remote Sensing Satellites: The Satellites with the Largest Number of Launches

Recoverable remote sensing satellites are the kind of satellites with the largest number of launches among the various different kinds of satellites that China developed in the 20th century. It is also these satellites that achieved the earliest development as it took the lead in entering the practical stage, reaching internationally advanced levels. The development and launch of recoverable remote sensing satellites was aimed at conducting earth observation and land and resources investigations.

In the mid-1960s, China set out to study the first recoverable satellite, for which Wang Xiji, a great scientist in the "two-bombs and one-satellite" programs and academician with CAS, acted as chief designer. At the beginning of 1966, under the leadership of Wang Xiji, experts started to work on the demonstration of the overall recoverable satellite program. When talking about the reasons for China's emphasis on recoverable satellites as the focus of space development, Wang Xiji recalled: "China needed to observe itself from the height of an orbit, where it could clearly see its territory, including surrounding oceans and coasts. It was difficult to achieve that by any other means at that time. In addition to earth observation, there was also another reason for development of recoverable satellites. At that time, the Soviet Union had been to space and the US was also working hard to study manned spaceflight. In consideration of the nation as a whole, we had to develop manned spaceflight, for which space recovery technology was fundamental."

Development of recoverable satellites necessitates breakthroughs in satellite attitude control, heat prevention in satellite re-entry, satellite recovery and other key technologies. Wang Xiji said: "We cannot adopt the principle of the minority being subordinate to the majority with regard to technological issues and, instead, we should respect objective rules and persevere in seeking truth from facts, as sometimes what the minority insists may be right." These words are still fresh in the memories of many people in the aerospace field.

On November 26th, 1975, China's first recoverable satellite was sent into space on LM-2 and safely returned to Earth three days later. With the successful development and launch of this satellite, China became the third nation in the world to have mastered recoverable satellite technology following the US and the Soviet Union. Globally, this technology was difficult to discover and utilise, and even today, there are only a few countries which have mastered this technology.

The first recoverable satellite had a blunt conical shape with a maximum diameter of 2.2 m, a height of 3 m and an internal volume of

7.6 cubic meters. The vehicle with a total mass of 1,790 kg, was divided into two major sections: the equipment module and the re-entry module. The satellite carried a visible light camera, which took photos of pre-determined zones in Chinese territory.

▶ First Recoverable Satellite

In the past 30-odd years, China has developed and launched more than 20 recoverable remote sensing satellites of five different models, including national land investigation satellites, remote sensing and mapping satellites and detailed land investigation satellites, all creating favorable social and economic benefits. Through developing these satellites, China has solved overall design, manufacture, large-scale experiment, launch, tracking, monitoring and control, recovery and other key technologies for recoverable remote sensing satellites. In particular, successful fulfillment of FSW-3 and FSW-4 (FSW being an abbreviation for *fanhui shi weixing* meaning recoverable satellites) missions has further matured and gradually developed the recoverable satellite platform and also greatly enhanced payload and other performance factors of recoverable satellites.

▶ Fengyun Satellite: "Our Own Meteorological Satellite"

Meteorological satellites are a type of applied satellites closely associated with the national economy and people's daily life. They can acquire information on atmospheric changes and effectively improve accuracy of weather forecasts; therefore, they play an irreplaceable role in disaster alleviation, natural calamity prevention and environmental protection. China began the development of meteorological satellites in the 1970s and has developed Fengyun 1 and FY-3 polar orbiting meteorological satellites and FY-2 geostationary meteorological satellite.

In January 1969, when Premier Zhou Enlai received representatives from the Central Weather Bureau and other units, he said that China should put an end to its backwardness by "developing our own weather satellites". In February 1970, CCCPC, the State Council and the Military Commission of the CCCPC assigned the mission of meteorological satellite development and, under overall arrangement and organization of the National Defense Science and Technology Commission, Shanghai organized relevant departments to engage in the development of polar orbiting meteorological satellites. In November 1977, the National Defense Science

▶ Satellite Cloud Picture Photographed by FY-2

and Technology Commission convened the meeting for first general feasibility plan for the meteorological satellite program to carry out technological coordination and demonstration for the overall plan and various major systems. At the meeting, the meteorological satellite program was codenamed 711, and the first polar orbiting meteorological satellite was dubbed FY-1.

The FY-1 meteorological satellite was a big project. On September 7th, 1988, China's first experimental polar orbiting meteorological satellite FY-1 was sent into space from the Taiyuan Satellite Launch Center on an LM-4A carrier rocket, making China the third country in the world with the ability to develop polar orbiting meteorological satellites and exercising great influence both at home and abroad.

The FY-1 meteorological satellite comprises seven major subsystems for meteorological remote sensing, image transmission, configuration, thermal control, power source, attitude control and surveillance and control, as well as onboard computers and antennae. The satellite adopted a hexahedron shape with its length, width and height being 1.4 m, 1.4 m and 1.2 m respectively and had solar wings installed on both sides. Weighing 750 kg and operating in orbit at an altitude of 901 km, FY-1 can acquire high quality cloud atlas and capture images on the frontal cloud system, extratropical cyclones, rainstorm cloud clusters, equatorial radiation and other weather systems.

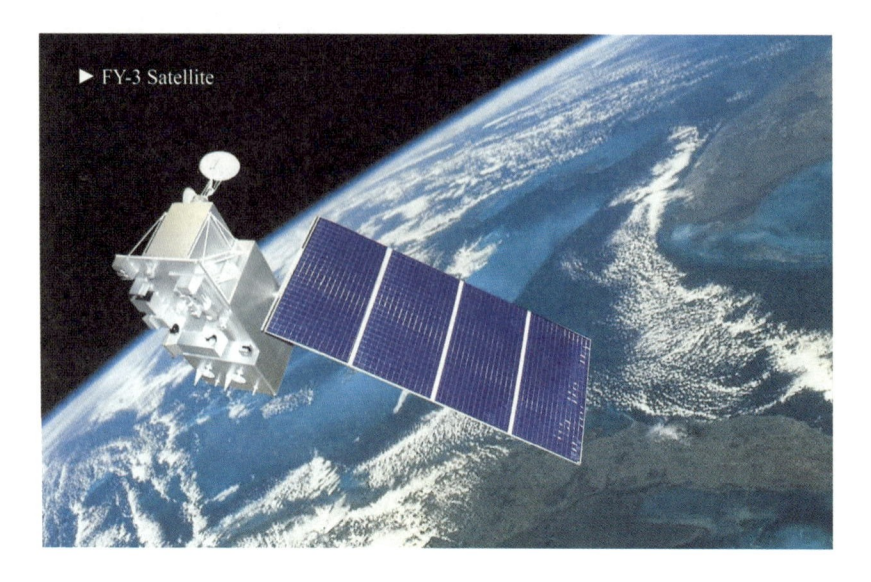

▶ FY-3 Satellite

On May 27th, 2008, China launched FY-3, a new generation of polar orbiting meteorological satellite, on LM-4C, achieving a new milestone in the exploration functions of China's meteorological satellites. For FY-3, the ground resolution of five of its channels reached 250 m, more than three times higher than that of FY-1.

FY-3 also used microwave remote sensors for the first time, with an exploration frequency as high as 183 GHz. These technological innovations greatly improved the exploration level of the satellite and enabled the satellite to render color and three-dimensional images and detect thickness of clouds with the results similar to those of a full body CAT scan.

From June 10th, 1997 to December 8th, 2006, China succeeded in launching 4 meteorological satellites in the FY-2 series. These satellites adopted a cylinder shape with a height of 1.6 m, a diameter of 2.1 m and a mass of 1,369 kg and had their surfaces covered with about 200,000 solar cells, which make the satellites especially suitable for monitoring the occurrence and development of disastrous severe convection weather systems which have a short life cycle but cause great harm.

▶ Resource Satellite: A Model of "South-South" Cooperation

As humankind's demand for natural resources strongly increases, the traditional natural resource exploration method can no longer meet the demand. In this regard, with continuous, rapid, global, regular, detailed and comprehensive observation capabilities, earth resource satellites can conveniently acquire images of and data from the Earth's surface, not only contributing to resource investigation for humankind but also playing a great role in disaster prevention and alleviation.

In the 1970s, China began the relevant research for the earth resource satellite program. However, as a result of a lack of financial support, the development work failed to make much progress. It was not until 1986 that the State Council approved the report of the Ministry of Aviation and Aerospace Industry on accelerating the development of aerospace technology and decided to provide necessary support to the development of three kinds of applied satellites (resource satellites being included in the three), marking the official beginning of China's resource satellite program.

Coincidentally, Brazil was also considering the development of resource satellites at that time. In 1987, leaders and experts with the Ministry of Aviation and Aerospace visited Brazil as the Brazilian government had expressed a great interest in the earth resource satellites which were being developed by China. After talks, both parties agreed to jointly develop earth resource satellites. Brazil and China were both developing countries with abundant resources. Although they adopt different political systems, they were brought together by common interests. On July 6th, 1988, in Beijing, the governments of the two countries signed "The Protocol Approved by the Government of the People's Republic of China and the Government of the Federative Republic of Brazil on Joint Research and Development of Earth

Resource Satellites".

According to the protocol, both parties would jointly develop and produce two earth resource satellites based on equality and mutual benefits and for the purpose of peaceful use of outer space. The China-Brazil earth resource satellite (CBERS) program represented a total investment of USD 150 million, including USD 100 million for satellite development and USD 50 million for satellite transportation and launch. 70% of the investment was assumed by China, and the remaining 30% by Brazil. This program was hailed as "a model of South-South cooperation in high technology".

On October 14th, 1999, China's first earth resource satellite CBERS-1 was successfully sent into sun synchronous orbit, and smoothly accomplished separation with the carrier rocket and expansion of its solar wings. Its successful launch put an end to the circumstances under which China had to purchase foreign satellite data due to not having its own land resource satellites.

CBERS-1 was a cuboid measuring 2 m in length, 1.8 m in width and 2.25 m in height and weighing 1,540 kg. It adopted a triple axis stabilization mode, had a design life of two years and operated in sun synchronous orbit at an altitude of 778 km. CBERS-1 went beyond some development stages of satellites of the same kind, as it was used immediately without launching any experimental satellite. Having a headstart and overcoming great technological difficulties, CBERS-1 was the satellite with more components and parts and a more complicated system than any other satellite in the history of China's satellite development. When commenting on CBERS-1, its chief designer Chen Yiyuan said: "As China's first generation of transmission satellites for earth observation, CBERS-1 was completely homegrown as the satellite was designed by China on the one hand, and all of its significant components were also made in China on the other, such as its momentum wheels, driving devices and expansion mechanism for solar wings, gyros, infrared horizon sensors, remote sensing cameras, high bit rate digital transmission, etc. It represents the highest level of China's remote sensing satellites and has bolstered the reputation of China. Its successful development marks China's development of remote sensing satellites having reached a new stage."

More than three years' operation of CBERS-1 in orbit saw its extensive use in agriculture, forestry, geology, energy, hydraulics, topography, petroleum and environmental protection. Chen Shupeng, an academician with CAS, has spoken highly of CBERS-1 for its wide application nationwide: "CBERS-1 data has its own features and advantages and we should bring into full play its role in economic development. In particular, we should make the utmost use of this data in the current development of the western regions...I believe that our resource satellites will make greater and better achievements."

Since CBERS-1, China has launched several resource satellites for land and resource investigation and monitoring, disaster prevention and alleviation, agriculture, forestry and hydraulics, ecological environment, urban planning and construction, transportation and major national projects.

▶ Ocean Satellite: the Remotely Operated "Scientific Manager" of the Oceans

Entering the 21st century, "marching towards the oceans" became a significant strategic objective in the economic development of many countries. China is a large ocean nation with a coastal area of 3 million square kilometers. With regards to ocean development and utilization, China is faced with increasingly serious resource and environmental problems, which not only affect the sustainable development of the marine economy and sustainable utilization of the oceans, but also directly relate to survival of humankind and their social development. Therefore, China's coastal areas and territorial seas appealled urgently for a "scientific manager" who was able to monitor the oceans from space and in a remote manner.

China's ocean satellite program divided ocean satellites into three series: ocean color and environment satellites (Haiyang-1 or HY-1,

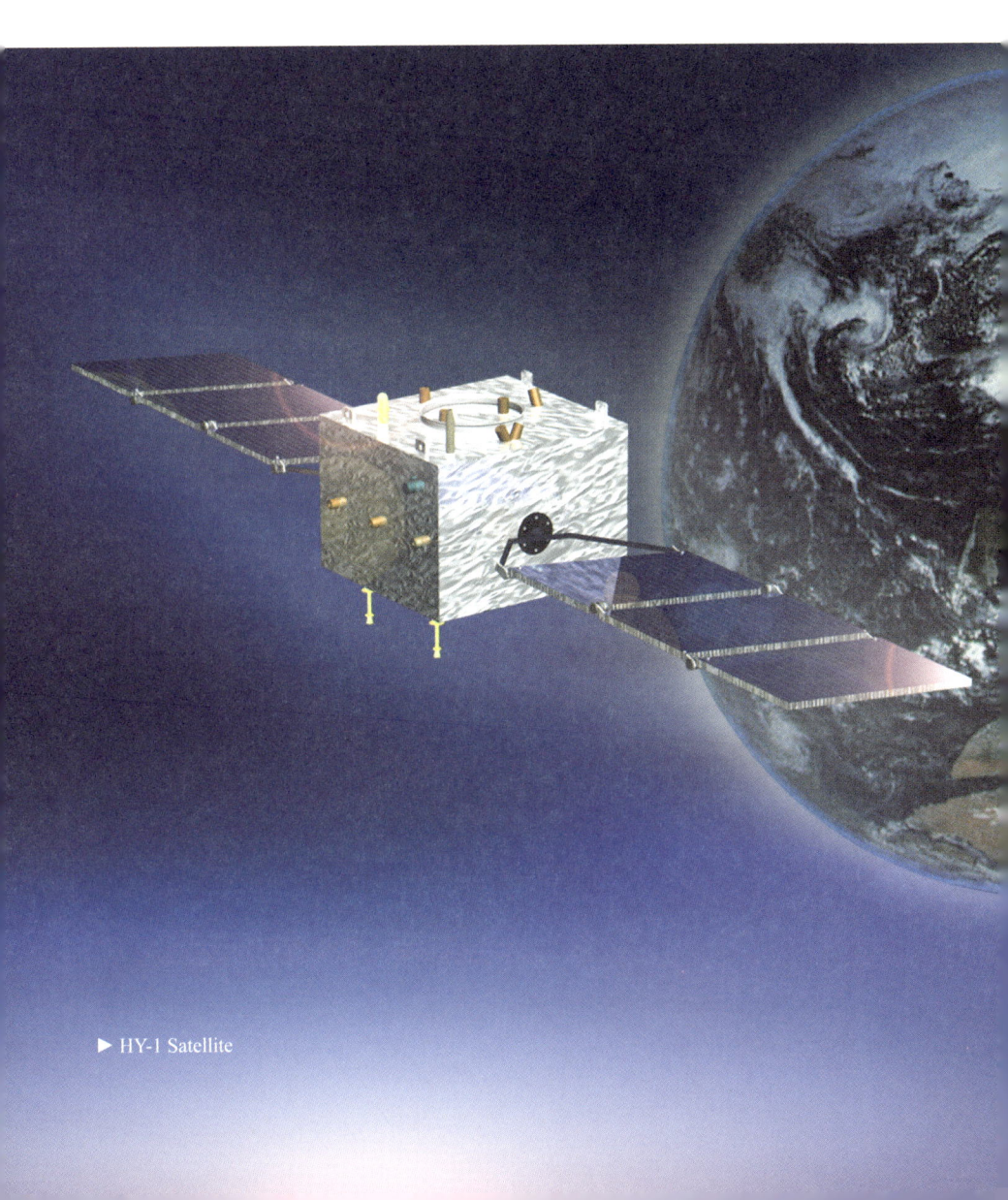

▶ HY-1 Satellite

haiyang literally meaning "ocean"), ocean dynamic environment satellites (HY-2) and ocean radar satellites (HY-3), in order to meet the ocean monitoring objectives of having a dedicated satellite, long life cycle and continuous and stable operation.

In the late 1980s, China used the FY-1 satellite to monitor China's ocean color for the first time. In 2001, the National Satellite Ocean Application Service (NSOAS) was founded, which has played an active role in building China's ocean satellites and their application systems. To further improve China's independent ocean monitoring system, China's first experimental service satellite for remote sensing and exploration of ocean color came into being.

On May 15th, 2002 and April 11th, 2007, China launched two HY-1 satellites, bringing an end to China's history of having no ocean satellites and representing the important progress that China had made in remote sensing application technology for ocean satellites. It also marked China's moving into the ranks of countries worldwide that have advanced ocean satellite remote sensing technology.

According to Zhang Yongwei, commander-in-chief and chief designer of the HY-1 satellite, development of the ocean color satellite was a great leap forward for China, as China achieved great progress after starting from scratch. HY-1 solved red tide observation and ocean right and interest protection issues as far as they related to the national economy and national security. It also played an important role in forecasting oceanic disasters and protecting national wealth and people's lives. At the same time, it also established the second ocean satellite application system following the national meteorological satellite program. This is a new and enduring market, and of great significance to the development of China's space technology and national economy.

On August 16th, 2011, the HY-2 satellite took off as China's first ocean dynamic environment monitoring satellite. It had a microwave scatterometer, a radar altimeter, a scanning microwave radiometer, a calibrated microwave radiometer and four microwave remote sensors. The satellite possessed all-weather, round-the-clock and all-world continuous exploration capability and was able to perform high-precision and synchronous measurement of multiple elements for oceans worldwide.

The HY-2 satellite also plays a vital role in ocean environment monitoring and forecasting, resource exploitation, maintenance of maritime rights and interests and scientific research. More specifically, the satellite can monitor storm tide, huge waves and other extreme ocean phenomena in a continuous and efficient manner to improve timeliness and effectiveness of oceanic disaster warning, identify and provide important ocean fishery information to offer technological support for the development of ocean fishery resources and acquire data on changes in the global sea level and polar ice caps to support studies on global climate changes.

▶ Environmental Satellite: To Perform Environmental and Disaster Monitoring and Forecasting

China is one of the countries around the world that suffers from some of the most severe natural disasters. Its vast territory, complex terrain and changeable climate ensure that its natural disasters always affect large areas, occur frequently and lead to heavy losses. It is a fact that 70% of Chinese cities and 50% of the population are situated in those regions that are susceptible to serious meteorological, seismic and geological hazards as well as other natural disasters.

On February 12th, 2003, the program to initiate a small satellite constellation for environment and disaster monitoring and forecasting obtained formal approval from the state. This is a brand new civilian satellite program following in the footsteps of meteorological, ocean and resource satellite programs. It is also China's first satellite monitoring system dedicated to disaster and environment monitoring. This also made China the first country in the world to use small satellite constellations for disaster prevention and alleviation. Construction of this satellite system was a significant milestone in the development of national environmental monitoring as it signified China's environmental monitoring entering the satellite application era.

On September 6th, 2008, Huanjing 1A (HJ-1A, *huanjing* literally meaning "environment") and Huanjing 1B (HJ-1B) were successfully launched on LM-2C from the Taiyuan Satellite Launch Center by having lift off of "two satellites with one single rocket" and placed in a preset orbit. HJ-1A and HJ-1B shared the same orbit and operated in quasi-sun synchronous orbit at an altitude of 650 km and an inclination of 98 degrees. The two satellites differed in a phase by 180 degrees and had a revisit cycle of two days after they formed a network. In addition, HJ-1A

▶ HJ-1A Satellite

▶ HJ-1B Satellite

▶ HJ-1C Satellite

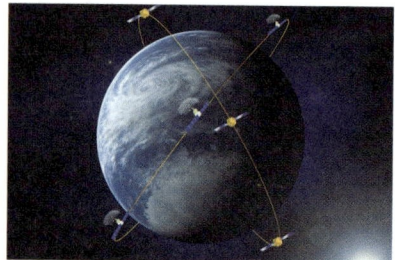

▶ Small Satellite Constellation for Environment and Disaster Monitoring and Forecasting

also undertook Asia Pacific multilateral cooperation tasks and carried Ka communications experiment transponders developed by Thailand.

On November 19th, 2012, HJ-1C was sent into space from the Taiyuan Satellite Launch Center. Weighing 890 kg and operating in sun synchronous orbit at an altitude of 500 km, HJ-1C has all-weather and around-the-clock operational capabilities and some penetrating power, which is particularly useful under adverse weather conditions.

▶ High Resolution Program: A Global Integrated Program Covering the Sky, Space and Earth

With the development of social economy, high resolution and high precision are undoubtedly the development objectives of the earth observation system. In 2007, China began to build its advanced atmospheric, land and ocean observation systems which consisted of the three observation platforms at earth, space and sky levels respectively. Currently, it has effectively formed a space-based earth observation system including four civilian satellite series for meteorological, oceanic, resource and environment and disaster alleviation purposes. It is predicted that by 2020, China's self-sufficiency rate of space data will have been increased to 60%-80%.

Nevertheless, China has not built a complete high resolution earth observation system, with its acquisition of high-resolution remote sensing data relying on the aerial remote sensing system and foreign high-resolution satellite systems. At the same time, China has a huge demand for earth observation data, especially for high spatial resolution data. Against this backdrop, it is becoming urgent for China to construct its own high-resolution earth observation system and build a global integrated all-weather and around-the-clock earth observation system featuring high space, high time and high spectral resolution and covering the sky, space and earth.

In 2006, China issued the "Guidelines for the National Medium- and Long-term Program for the Science and Technology Development (2006-2020)", which explicitly details the implementation of 16 major scientific and technological programs and major science and technology infrastructure programs, which include the major program of the high-resolution earth observation system ("high-resolution program").

In the same year, China started its demonstration of the implementation plan for the high-resolution program. On May 12th, 2010, the high-resolution program entered the comprehensive initiation and implementation stage. By 2020, China will have developed its own independent high-resolution earth observation system. Implementation of this program will meet the needs of national economic and social development. It will also be of immense significance in promoting the construction of China's space infrastructure, in cultivating satellite application enterprise clusters and an industry chain and in driving the development of satellite application and strategic emerging industries.

On April 26th, 2013, Gaofen 1 (GF-1, gaofen literally meaning "high

▶ GF-1 Satellite

| Chinese Satellites in Space |

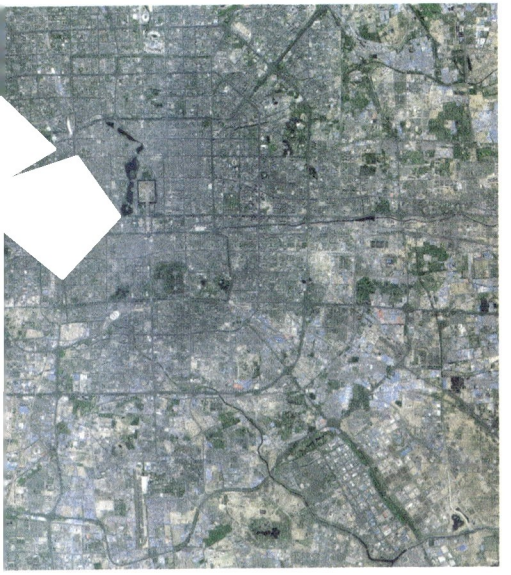

▶ Beijing Images Obtained by GF-1 Satellite   ▶ Shanghai Images Obtained by GF-1 Satellite

resolution"), the first satellite for the high-resolution program, was successfully launched. Bai Zhaoguang, commander-in-chief and chief designer of GF-1, introduced it by saying: "GF-1 is China's first low-orbit earth observation satellite with a design and assessment life longer than five years, which has undoubtedly set a new record." With its positioning accuracy for images reaching dozens of meters and attitude stability meeting 5/10,000 degree per second, GF-1 ensures image quality and achieves a perfect combination of spatial resolution and temporal resolution. Moreover, since GF-1 adopts lightweight compact design and new materials, it enjoys a high payload ratio.

On June 6th, 2013, China released the first batch of 13 images acquired by GF-1, which covered the four cities of Beijing, Shanghai, Yinchuan and Datong. This batch of images, distinct in gradation, clear in image and containing much information, indicates that all systems of GF-1 operate well and meet the design requirements and that GF-1 has accomplished the goal of providing an accurate service in national land investigation, environment, agriculture and other fields and greatly improved the overall observation capability of China's earth observation satellites.

Communications Satellites as Bridges for Space Information

China set about developing communications satellites in the 1970s and 1980s. Now, after three decades of effort, it has developed a communications satellite series with extensive business. By insisting on independent innovation and development, targeting internationally advanced levels and combining scientific research and practice as its principles and through development and application of three generations of practical communications satellites, China has successfully shifted the focus of its communications satellite development from exploration to practice, from experimentation to application and from the Chinese market to the international market. It has become one of the few countries in the world to be able to develop high-capacity communications satellites, greatly contributing to China's development of its social economy and construction of its national defense.

▶ DFH-2

In 1975, the National Planning Commission and the National Defense Science and Technology Commission jointly presented the "Report on Development of China's Satellite Communications" and obtained approval from the central authorities. As a result, the satellite communications program was officially included in the state plan. In

Chinese Satellites in Space

DFH-2 Satellite

May of that year, at the planning symposium organized by CAST, all experts present believed that China should launch its communications satellites as soon as possible and occupy advantageous orbital slots, which would benefit development of China's geostationary satellites and thus be an urgent strategic task. Zhang Aiping, Director of the National Defense Science and Technology Commission, pointed out that "We should take communications satellites as the focus of our work and concentrate our efforts on their development, as they are of great political and economical value."

On April 8th, 1984, China's first geostationary experimental communications satellite DFH-2 was successfully launched. The satellite adopted a cylinder shape with a diameter of 2.1 m, a height of 1.6 m, a weight of 920 kg, and a service life of 3 years. With two C band transponders and global horns for both receiving and transmitting purposes on board, the satellite can provide all-weather and around-the-clock communications services, including television, telephone and radio, therefore assuming some domestic communications tasks. With successful operation of DFH-2, China became the fifth nation in the world to have independently launched a geostationary satellite, signifying that China had made its first step forward in the development of communications satellites.

▶ DFH-3 Satellite Platform

With the rapid development of China's satellite communications cause, DFH-2 could no longer satisfy the needs of its users. Therefore, China embarked on the development of the medium-capacity communications satellite, DFH-3, in 1986.

On May 12th, 1997, DFH-3 was successfully launched, marking China's satellite communications technology moving forward into a new stage. DFH-3 is box-shaped with a length of 2.22 m, a width of 1.72 m and a height of 2 m, and is equipped with large extendable solar wings and antennae and with a diameter of 2 m. The satellite adopts a configuration of three modules: the communications module, the propulsion module and the service module. The platform also consists of six subsystems for control, power source, surveillance and control, propulsion, thermal control and structure. By the end of 2012, 27 satellites had been sent into space based on DFH-3 satellite platform.

▶ China's First Relay Satellite Based on DFH-3 Satellite Platform: Tianlian 1

▶ DFH-3 Satellite Platform

▶ DFH-4 Satellite Platform

In the 1990s, China's demand for satellite communications came to a climax, with its satellite level and operation scale not able to meet its business needs. As a result, many satellite companies formed in order to purchase foreign communications satellites and deal in satellite communications, all of which was very common during this period. At the same time, foreign communications satellite manufacturers also seized the opportunity to expand their shares on the Chinese market.

Under such grim situations, China had to develop its own large-capacity and long-life communications satellite platform. Therefore, China concentrated its efforts on key technological breakthroughs in the large geostationary satellite platform during the "Ninth Five-year Plan" period and gave official approval for the establishment of the DFH-4 satellite platform program in October 2001.

When talking about difficulties in the development of the platform, Zhou Zhicheng, chief designer of the DFH-4 satellite platform, said: "The DFH-4 satellite platform is a commercial communications satellite platform and there is no previous experience as a reference point so we have to start from scratch. That is a very challenging task as we must develop an international first-class communications satellite platform with an output power of 10,000 W, a payload of 600-800 kg, a service life of 15 years and able to carry 52 transponders, based on the current DFH-3 platform with an output power of 1,700 W, a payload of 200 kg, a service life of 8 years and only able to carry 24 transponders… For the sake of the survival of our national industry, we must accomplish this task."

On May 14th, 2007, China succeeded in launching NigComSat-1, the first communications satellite developed on the DFH-4 satellite platform. With successful development of the DFH-4 satellite platform, China mastered many key technologies, achieved wide-ranging and groundbreaking results with independent proprietary intellectual property rights, promoted great steps forward in the development of China's space technology and spurred technological advances and product upgrades in related basic industries for components, parts and raw materials. Meanwhile, it also put an end to the inactivity which allowed China to be under the control of foreign countries as a result of its long-term dependence on the import of communications satellites.

Currently, China has developed seven communication satellites based on the DFH-4 satellite platform and successfully put them into space, which are stably operating in orbit, and there are another dozen of these satellites under development. This has also put China in the league of the world's leading countries that are able to develop large communications satellites.

Beidou Navigation Satellites

On December 27th, 2012, the official version of Beidou Navigation Satellite System Signal in Space Interface Control Document was released. Beidou Navigation Satellite System then formally provided passive positioning, navigation and timing services the Beidou navigation services covering most Asia-Pacific region. Beidou Navigation Satellite System, together with the U.S. Global Positioning System, the Russian GLONASS and the E.U. Galileo positioning system, was confirmed by the United Nations as one of the core suppliers of global satellite navigation systems.

For a modern power that has stepped into the information era, a satellite navigation and positioning system has become an important national information infrastructure and strategic facility. It is also a significant manifestation of the comprehensive national strength of a country. The United States began to develop and deploy GPS in the 1970s, and has established its leading position in the sector. China must build its own satellite navigation and positioning system, which has long been the cherished wish and pursuit of several generations of Chinese people.

▶ Beidou Navigation Experimental Satellite

To seek a simple and practical system with a few satellites, Chen Fangyun, a renowned scientist from the "two-bombs and one-satellite" programs, proposed the concept of using geostationary satellites to construct China's own satellite navigation and positioning system in 1983. According to Chen, in view of China's national circumstances at that time, China could develop a double-satellite positioning system involving relatively simple technologies and requiring lower costs.

In 1985, at the national measurement technology symposium convened by Nanjing Purple Hills Observatory, Academician Chen Fangyun put forward the suggestion of building a double-satellite positioning system. "Science and engineering are two different things, and there is a huge gap between them. Since even foreign countries have not done this before, it is impossible for us to achieve this in consideration of our current technological level." In the face of such doubts, Academician Chen Fangyun solemnly declared that: "A double-satellite positioning system is scientifically possible and feasible."

In 1986, with implementation of the national "863" program, the progress of building the satellite navigation and positioning system was also accelerated. In the same year, China approved the establishment of the program and the pre-research. In 1993, the double-satellite positioning system was incorporated in the national "Ninth Five-year Plan". In January 1994, the Beidou satellite navigation experimental system program (beidou literally meaning "Big Dipper") and the pre-research obtained official approval, the program being codenamed "Beidou 1" (BD-1) with Academician Sun Jiadong as the chief designer.

There was no precedent for double-satellite positioning in the world, let alone drawings and materials for learning and reference. Each step forward in the development of the

▶ Two "Beidou" Satellites Installed atop One Rocket for Launch

program is still vivid in the memories of the experts that were engaged in the program. Tan Shusen, deputy chief designer of the Beidou program system, once said enthusiastically: "Although China started developing its satellite navigation and positioning system rather late, we finally rank among the top three. That is because we have explored an independent innovation road with Chinese characteristics."

With a series of successive key technological breakthroughs, the satellite navigation and positioning system was finally successfully developed. The Beidou navigation experimental satellite was developed on the DFH-3 satellite platform and equipped with navigation payload. On October 31st and December 21st, 2000, two Beidou navigation experimental satellites were sent into space, forming a network and building the initial Beidou satellite navigation experimental system. On May 25th, 2003, the third Beidou navigation experimental satellite was placed into orbit, further strengthening performance of the Beidou satellite navigation experimental system. Formal operation of the system enabled China to be the third country in the world to have its own satellite navigation and positioning system, following the US and Russia.

Combining the three major functions of satellite navigation, message communication and high precision timing, the Beidou satellite

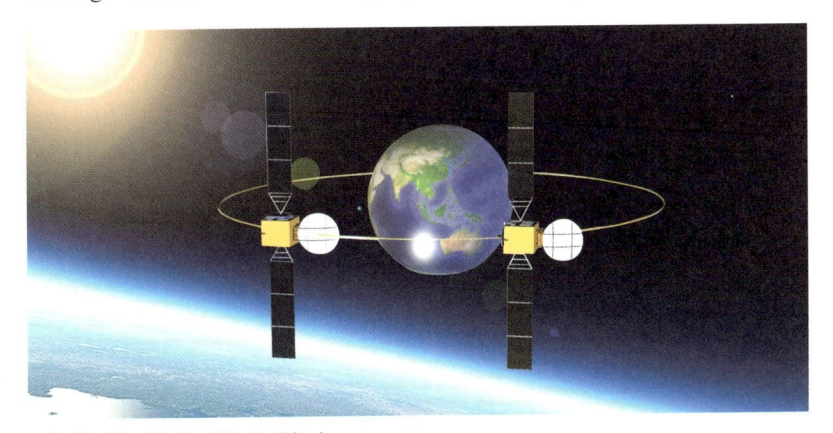

▶ Beidou Double-Satellite Positioning

navigation experimental system can not only solve positioning problems such as "Where am I?" and perceiving problems such as "Where are you?", but it can also carry out efficient and convenient information transfer between "you" and "me". When commenting on the Beidou system, Zhao Kangning, Deputy Director of the Chinese Satellite Navigation and Positioning Applications Management Center, said: "The Beidou navigation satellite is an excellent representative for being 'made in China' and it stands greatly as an independent and innovative creation of the Chinese nation."

The Beidou satellite navigation experimental system is part of China's information infrastructure and is a strategic facility of great national importance, and it has brought about remarkable economic and social benefits in hydraulics and hydroelectricity, marine fishery, transportation, weather forecast, land surveying and mapping, disaster alleviation and relief and public security.

On May 12th, 2008, the Wenchuan earthquake led to the disruption of transport, water supply and communications. On an emergency basis, the Chinese Satellite Navigation and Positioning Emergency Management Center equipped the rescue forces with more than 1,000 Beidou user terminals, which played a crucial role in the earthquake relief work, as it transmitted over 40,000 disaster messages to the headquarters and performed positioning services over 130,000 times. This provided a great deal of information for decision-making and command from headquarters and saved time for success in combat against the earthquake.

▶ Beidou Navigation Satellite

Current and future trends in satellite navigation systems are to achieve global coverage, compatible with other navigation systems. Therefore China launched the Beidou Navigation Satellite System project in 2004.

| Chinese Satellites in Space |

The Beidou system is comprised of three major components: space constellation, ground control segment and user terminals. It is planned to begin serving global customers upon its completion in 2020. The space constellation consisted of 5 geostationary orbit (GEO) satellites, 27 medium-Earth orbit (MEO) satellites and 3 inclined geosynchronous orbit (IGSO) satellites; The ground control segment includes a number of master control stations, injection stations and monitoring stations. The user terminals include various Beidou user terminals, and terminals compatible with other navigation satellite systems to meet different application requirements from different fields and industries. In the short span of six years from 2007 to 2012, China launched 16 Beidou navigation satellites to form a satellite network, which began to provide services for China and its surrounding regions at the end of 2012. The

▶ "Beidou" Geostationary Satellite   ▶ Beidou Non-geostationary Satellite

Beidou satellite navigation system program saw the Chinese aerospace industry's first attempt at batch production. It was very challenging for a development team to manufacture a dozen of satellites according to a set of drawings, under the same technological conditions and within a short time. To ensure that Beidou satellite constellations could achieve networking in a short time, the development team had to stick to the development process of "once-through design, batch production, running testing, intense launch and rapid networking". This showed that China's aerospace research and development capabilities and management level had reached new heights.

The Beidou satellite navigation system officially provides regional services, speeding up satellite application industrialization led by the Beidou demonstration program. Now, China's satellite navigation enterprises have developed Beidou chips, modules and terminals with proprietary intellectual property rights and navigators with Beidou satellite navigation functions have been put into trial use. It is believed that the Beidou satellite navigation system will play more vital roles in various fields of China's social economy in the near future.

Experimental Satellites for Scientific and Technological Development

In order to perform preliminary experiments on new technologies, of which some tasks in space urgently required, and carry out space environment exploration and space science research, China began to develop science and technology experiment satellites in the initial stage of its satellite development. Since the 1970s, China has developed and launched the Practice series of satellites and two satellites under the "Geospace Double Star Exploration Program".

▶ **Practice Series of Satellites**

In 1968, when developing the DFH-1 satellite, academician Sun Jiadong and other researchers also put forward the idea of initiating the second satellite program which focused on experimentation of long-life power supply systems and carried out program design and key component testing. In May 1970, encouraged by the successful launch of DFH-1, the National Defense Science and Technology Commission reviewed and determined the overall plan for the Practice 1 satellite program. Based on the DFH-1, the Practice 1 program added eight space technology experiment and space exploration projects. The greatest difference between Practice 1 and DFH-1 lies in that the former is furnished with a long-term power source system, a long-term thermal control system and a long-term remote monitoring system.

▶ Comparison of Space Small Red Beans and Peas (Right of Each Set of Groups) with Common Species

On March 3rd, 1971, China's second man-made satellite, Practice 1, was shot into space. Practice 1 operated in orbit for more than eight years, during which it conducted experiments for the silicon solar cell power supply system, the active thermal control system and other key technologies for long-life applied satellites. On September 20th, 1981, China first launched, on one single rocket, three satellites, which were Practice 2 satellite constellations made up of Practice 2, Practice 2A and Practice 2B. Through this launch, China solved the difficult technological problem of multi-satellite separation, and became the third nation in the world to have mastered the technology of sending multiple satellites into orbit with a single rocket. At the same time, the satellites also acquired a large quantity of space environment exploration data of great scientific value.

On May 10th, 1999, China sent Practice 5 into space, which was also China's first small satellite by modern standards.

On September 9th, 2006, Practice 8 was successfully launched, which was China's first recoverable science and technology experiment satellite dedicated to space breeding research. This satellite brought more than 2,000 kinds of vegetable, fruit, grain, and cotton seeds with a total mass of 215 kg into space. Scientists hoped to probe the effect of weightlessness on seed germination through scientific experiments on Practice 8, thus exploring and mastering the laws of space breeding.

On October 14th, 2012, Practice 9A and Practice 9B were sent into space from the Taiyuan Satellite Launch Center by means of "launching two satellites on a single rocket". These two satellites, which have enhanced the domestication ability of China's aerospace products, are the first in a series of civilian satellites designed for technological experimentation. Images and data acquired by these satellites have been widely used in land resource investigation and monitoring, agriculture, forestry, hydraulics, urban and rural construction, environmental protection, disaster prevention and alleviation, etc., and meet the urgent need of its users for high-resolution data.

▶ Geospace Double Star Exploration Program

Geospace Double Star Exploration Program (from now on referred to as the "Double Star Program") was a program officially proposed by Chinese scientists in April 1997 and was organized by the CAS academician Liu Zhenxing. The program, officially started in February 2001, was designed to investigate the trigger mechanism of various disturbances as a result of solar activities and their regularity.

The Double Star Program is an innovative program in China's space science exploration, capturing the interest and attention of physics' international space community. In July 2001, CNSA and the European Space Agency (ESA) entered into a cooperation agreement in relation to the Double Star Program. Therefore, the Double Star Program saw China's first high-level, all-round, significant and equal cooperation with developed countries and its own advanced space exploration program.

▶ TC-1 and TC-2 Satellites

On December 30th, 2003 and July 25th, 2004, China launched Tance 1 (TC-1 tance literally meaning "exploration") and TC-2 satellites respectively. Developed on the CAST 968 platform, these two satellites both weighed 330 kg, adopted an attitude spin stabilization mode and looked cylindrical in shape with a diameter of 2.1 m and a height of 1.4 m.

The "Double Stars" of China and the four satellites under the "cluster" mission of ESA cooperated with each other to perform the simultaneous exploration of different regions. The cooperation also first put into action the six-point three-dimensional space exploration of Geospace in the human history. Scientific data acquired by them also strongly supported research on space physics and had a great influence internationally.

Chinese Satellites in Global Market

After entering the 21st century, China has seen its satellite development technology continuously improve and, in particular, the development of its common platform technology for satellites has leapt forward, which has paved the way for China's entry into the international market.

While developing DFH-4 satellite platform, CAST also began to open up its international commercial communications satellite market. In May 2004, the news came that Nigeria's communications satellite was calling for bids. Despite fierce competition in the international commercial communications satellite market, China also participated in the bidding. At the same time, another 21 companies from the US, France, the UK, Italy, Israel, etc. were also eager to enter the bidding. In the end, China won the bid with the good compatibility, high efficiency, excellent technological coordination, ever-increasing reliability and quality service of both its satellites and its rockets. On December 15th of that year, China Great Wall Industry Corporation and Nigeria's National Space Research and Development Agency signed the contract on the NigComSat communications satellite program in Abuja.

In May 2007, NigComSat 1 was successfully launched, and it was delivered in July. It marked China's first satellite export and opened the door for China to the international commercial communications

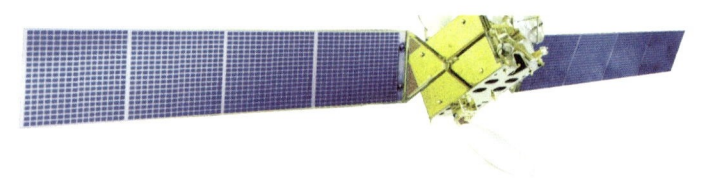

▶ NigComSat 1

satellite market. By the end of 2012, export communications satellite developed on the DFH-4 satellite platform had won orders from nine foreign customers from Pakistan, Laos, Myanmar and Indonesia in Asia, Venezuela and Bolivia in South America, Nigeria and Congo-Kinshasa in Africa and Belarus in Europe.

In addition to communications satellites, China has also realized export of its remote sensing satellites. In May 2011, China and Venezuela signed the contract on the Venezuelan Remote Sensing Satellite 1 (VRSS-1) program. On September 29th, 2012, the VRSS-1, developed on the established CAST 2000 small satellite platform, was successfully launched into space, which sent back the first remote sensing image of the territory of Venezuela on October 1st.

▶ Venezuelan Remote Sensing Satellite

On the road of internationalization, China's aerospace industry has gradually formed a satellite manufacture system geared to international standards and has further enhanced China's independent innovation capability and international competitiveness in satellite development.

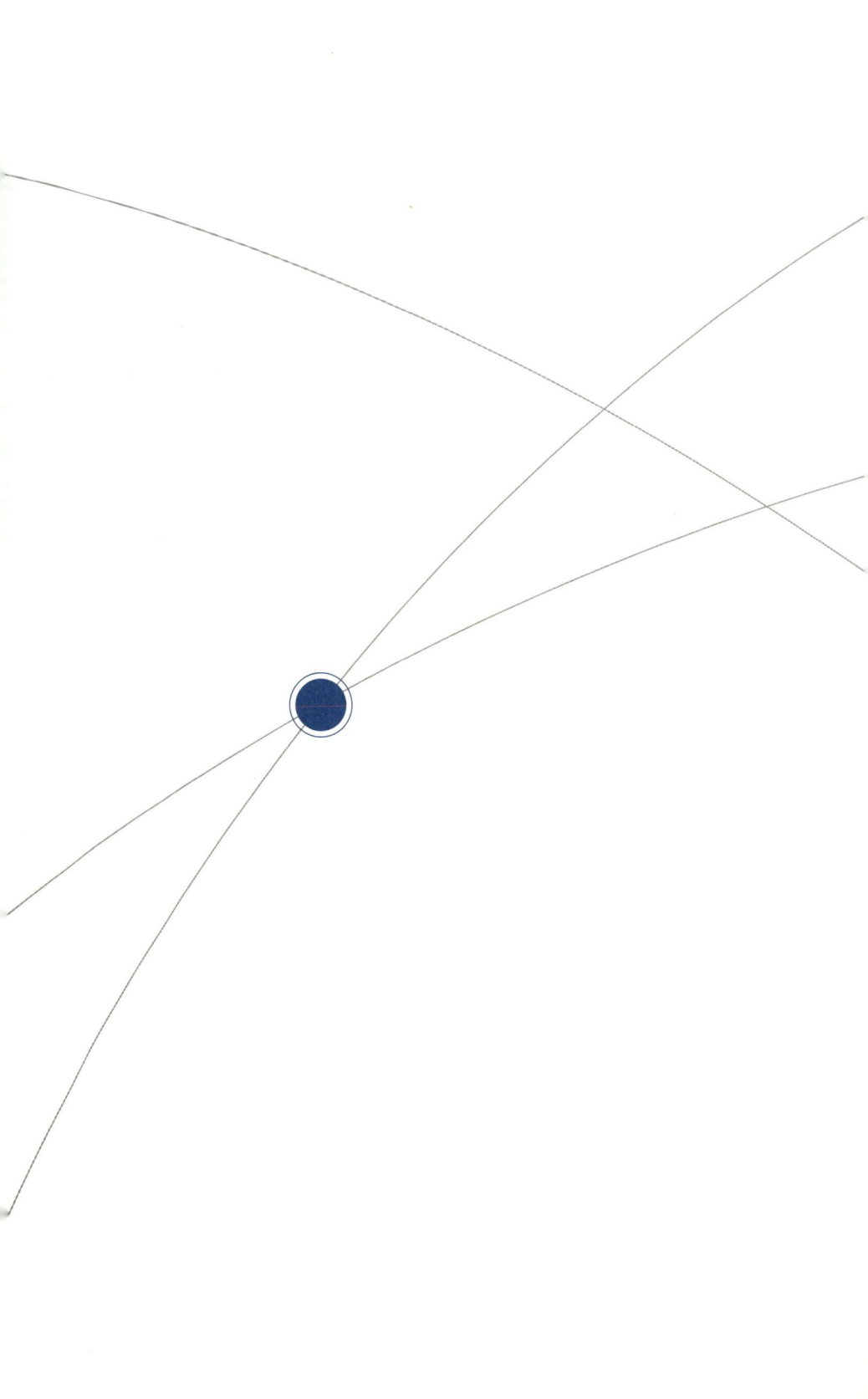

4

The Shenzhou Spacecraft: Chinese Nation's Millennium Dream of Spaceflight Comes True

On November 20th, 1999, China successfully launched and recovered the first unmanned experimental spacecraft, Shenzhou 1 (shenzhou literally meaning "divine vessel"), on LM-2F from the Jiuquan Satellite Launch Center. In the subsequent three years, China succeeded in lifting off and recovering Shenzhou 2, Shenzhou 3 and Shenzhou 4 unmanned experimental spacecraft. Launches of these four unmanned experimental spacecraft were aimed at ensuring high safety and reliability of the spacecraft and their success showed that China had fully mastered various manned spacecraft technologies and laid a solid foundation for birth of the first manned spacecraft.

On October 15th, 2003, China's first manned spacecraft carrying China's first astronaut traveled to space and returned to Earth safely. With this feat, China became the third nation in the world with the capability to independently carry out manned spaceflight activities, fulfilling its millennium dream of spaceflight.

In the decade from 2003 to 2013, China launched, in succession, five manned spacecraft, one unmanned spacecraft and one target spacecraft. It also realized a series of unprecedented and important breakthroughs, including its first astronaut travelling to space, multi-manned and multi-day spaceflight, its first space walk, its first space rendezvous and docking, its first female astronaut travelling to space and its first applied spaceflight. During this period, China accomplished tasks for the first step and the first phase of the second step of the "three-step" strategy for the manned spaceflight program, paving the way for China's construction of a space station by 2020.

China's Self-developed Flying Vehicle: The Shenzhou Spacecraft

The Shenzhou spacecrafts are China's self-developed manned spacecraft weighing 8 tons, measuring 9 m in height and 2.8 m in maximum diameter and consisting of three modules: the re-entry module, the orbital module and the service module. Among them, the re-entry module and the orbital module are sealed capsules where astronauts work and live. In addition, the front of the spacecraft is equipped with a rendezvous and docking system to enable docking with other space vehicles.

The re-entry module of the Shenzhou spacecraft is located in the middle section and serves as the command and control center with seating for astronauts. During launch, entry into orbit and reentry of the spacecraft, astronauts are seated in the re-entry module. The re-entry module is also the only portion of Shenzhou spacecraft which returns to Earth. It resembles a bell shape with its diameter and length both being about 2.5 m and provides a relatively spacious and comfortable cabin environment even with three astronauts on board. The re-entry module is connected through a hatch on top of the orbital module. When spacecraft fly in space, astronauts can go to the orbital module through this hatch.

The layout of the re-entry module centers around the seating for astronauts. The three seats adopt a fan-shaped arrangement and have instrument panels, cameras, search lights, pilot levers and other common equipment next to them, parachutes for landing upon re-entry above them and thermal control, propulsion, crew, monitoring, control and communications, environmental control and life support subsystems below them. In the side bulkheads of the re-entry module, there are two round windows, with one for astronauts to observe scenes outside the window and the other for astronauts to operate optical sights, understand ground situations and pilot the spacecraft properly.

The orbital module of Shenzhou spacecraft is situated in front of the re-entry module and is cylinder-shaped with a diameter of 2.2 m and a length of 2.8 m. It is the living module for astronauts. The back of the orbital module is connected to the re-entry module. During spaceflights, astronauts have meals and rest in this module, where food, drinking water and other essentials are provided. After accomplishment of their missions, the Shenzhou spacecraft leave their orbital modules in orbit for various scientific experiments. The orbital module is installed with two solar wings, which provide the module with power, just like wings of a bird. Since the astronauts of Shenzhou 7 must perform spacewalks, the orbital module was transformed into an airlock module, from which astronauts step into vast outer space.

The service module follows the re-entry module. It is a cylinder-

▶ Shenzhou 1

▶ Reentry Module of Shenzhou Spacecraft Landing Safely

shaped and non-sealed equipment module with a maximum diameter of 2.8 m and a length of 3 m. As the service module is mainly designed to provide spacecraft with power for their re-entry, it is also installed with two solar wings on both sides.

Shenzhou 5 Mission: First Manned Spaceflight

At 9:00 am on October 15th, 2003, China's first manned spacecraft Shenzhou 5 sent China's first astronaut, Yang Liwei, into space. When the spacecraft was flying around the Earth for the first time, Yang Liwei reported to the ground command center: "The flight is normal, and I feel good." This is the first voice that the intelligent and brave Chinese people heard from remote outer space.

After that, Yang Liwei "floated" to the window and caught sight of the vision outside of the window: the solar wings of the spacecraft were shining in the sun, under which there was the "cradle" of the humankind, the Earth, with its azure edge seemingly set in a golden rim and under thin clouds, the long coastlines between the land and the oceans were

▶ Shenzhou 5

still visible. Upon seeing that, Yang Liwei pressed the shutter in great excitement to record the historic moment. This was also the first time for the Chinese people to observe the Earth from space.

When the Shenzhou spacecraft flew around the Earth for the seventh time, Yang Liwei showed the world the Chinese national flag and the United Nations flag and said, in space, 343 km far away from the Earth: "Greetings to people all across the world, and to all colleagues engaged in space work! Thanks for the care from all the people of my motherland!"

When the Shenzhou 5 spacecraft successfully completed its mission, its chief designer Qi Faren was already more than 70 years old. When talking about the success of China's first manned spaceflight, he simply said: "In the eyes of others, history has been created by us, but as a matter of fact, we stood on the shoulders of our predecessors. Take a hungry man for example. With the first piece of steamed bread, he still feels hungry. After having the second piece, he doesn't feel so hungry.

After finishing the third piece, he has had enough. It appears that the third piece of steamed bread works, but how can the man feel full without the first two pieces." It was the selfless devotion of the older generation of scientists such as Qian Xuesen, Ren Xinmin and Sun Jiadong that brought about the final success of China's manned spaceflight.

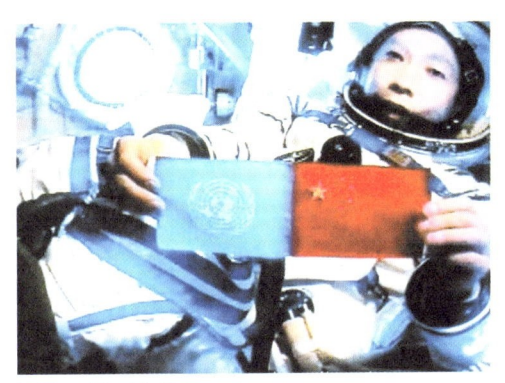

▶ Yang Liwei in Space

The Shenzhou 5 spacecraft flew around the Earth 14 times in a flight lasting 21 hours and 23 minutes and covering 600,000 km. With the accomplishment of this mission, China achieved many brilliant results: success of the maiden spaceflight showed that the overall spacecraft program and all system

▶ Shenzhou 5 Spacecraft and Yang Liwei

schemes were correct and that the spacecraft was fully qualified for manned spaceflight missions; during the entire spaceflight, all working and living conditions provided by the spacecraft to the astronaut met the requirements of the astronaut system; the astronaut and all recovered payload returned to Earth safely after completing the spaceflight mission; and the re-entry module remained intact. The Shenzhou spacecraft's success in China's first manned spaceflight marked a historic breakthrough in China's manned spaceflight program and inaugurated a new era for development of China's aerospace industry.

Shenzhou 6 Mission: First Multi-manned and Multi-day Spaceflight

At 9:00 am on October 12th, 2005, the Shenzhou 6 spacecraft traveled to space with a crew of Fei Junlong and Nie Haisheng. The commencement of the second step in China's manned spaceflight program was a connecting link between preceding and following spacecraft and so Shenzhou 6 had three major missions: to make further breakthroughs in basic space technologies, to continue scientific

▶ Shenzhou 6

experiments in space and to keep assessing and improving performance of other systems of the program.

To meet the needs of multi-manned and multi-day spaceflight, scientists and technicians improved and modified the Shenzhou 6 spacecraft by removing the additional section of the spacecraft, installing scientific experiment instruments and living facilities in the orbital module and equipping it with sufficient necessities for life. The latter of these was achieved by adding a hatch between the re-entry module and the orbital module to ensure safe movement of the astronauts between the two modules and adopting a people-oriented interior design for the two astronauts, creating a rather comfortable working and living environment.

The Shenzhou 6 spacecraft had a higher safety performance and a stricter flight control system. For the spacecraft, more than 150 on-orbit fault modes and their countermeasures were developed and, in case of any serious fault, the spacecraft could initiate emergency re-entry every time it flew around the Earth. Shenzhou 6 was launched on the LM-2F rocket, which was more reliable, since the Shenzhou 6 had an image real-time measurement system installed on board and a heater insurance system added to its escape tower and solved the vibration reduction problem, safeguarding smooth flight in the process of a launch.

▶ Shenzhou 6 Astronauts Having Training in Simulation Module

During the spaceflight, the astronauts onboard Shenzhou 6 opened the hatch of the re-entry module and entered into the orbital module, where they took off their space suits and changed into work clothes to carry out scientific experiments in space. This was China's first scientific experiment in space and with human participation. In the orbital module, the astronaut Fei Junlong also performed four forward rolls and took digital images in space.

After the Shenzhou 6 spacecraft had flown around the Earth 77 times, lasting about five days and covering 3,250,000 km, it safely returned to Earth with the two astronauts on October 17th, 2005, marking the great success of the spaceflight mission. Moving from Shenzhou 5 to Shenzhou 6 was never meant to simply be about an increase in number. Instead, it marked a giant step forward for China's manned spaceflight cause. With an increase in the crew of one member to two members, addition to the travel time of one day to several days and expansion of the astronaut activity scope from the re-entry module to the orbital module, China's manned spaceflight program had entered a new stage with direct human participation in scientific experiments in space.

Shenzhou 7 Mission: First Spacewalk

On September 25th, 2008, the Shenzhou 7 spacecraft carried the three-member crew of Zhai Zhigang, Liu Boming and Jing Haipeng into space in China's third human spaceflight. During the mission, a Chinese astronaut carried out his debut spacewalk.

Spacewalk, also termed extra-vehicular activity (EVA), refers to an astronaut leaving the spacecraft and entering into space. It

The Shenzhou Spacecraft:
Chinese Nation's Millennium Dream of Spaceflight Comes True

▶ Shenzhou 7

is a key technology in the manned spaceflight program and also a significant means for on-orbit installation of large equipment, scientific experimentation, satellite release and space vehicle examination and maintenance in the manned spaceflight program. To make a Chinese astronaut's first spacewalk possible, China made a series of technological breakthroughs: construction of the largest weightless water pool in Asia to train the crew for the spacewalk, transformation of the orbital module to the airlock module to create the conditions necessary for the astronaut to leave and return to the spacecraft and development of space suits to ensure the astronaut's safety in EVA. Moreover, China also launched the Yuanwang 6 space tracking ship and the Tianlian 1 data relay satellite, greatly increasing the surveillance, control and communications coverage of the spacecraft.

To accomplish the spacewalk mission, Shenzhou 7 spacecraft not only inherited the established technology of previous spacecraft but also adopted a new orbital module design. Compared with the orbital module of Shenzhou

▶ Shenzhou 7 Crew

China's Journey to Space: From Dream to Reality

6, the improved orbital module of Shenzhou 7 had the pair of solar wings removed, several spheroidal gas cylinders installed on top and a miniaturized satellite attached to it. In terms of the internal structure, the orbital module was equipped with a repressurized gas cylinder, two extravehicular space suits, pressure relief and recovery control equipment, an extravehicular safeguard control panel and other onboard supporting equipment, and furnished with sleeping bags, food heating facility, personal living supplies, personal sanitary installations and other living facilities.

On September 27th, 2008, after getting ready in the orbital module, the astronaut Zhai Zhigang made the "first step" towards space at 16:48. Zhai Zhigang, wearing a feitian (feitian literally meaning "Apsaras", a female spirit of the clouds and waters in Hindu and Buddhist mythology) space suit, slipped out of the orbital module in a head-first position and

▶ Training of Shenzhou 7 Astronauts before Travel to Space

reported to the camera on the service module: "Shenzhou 7 has left the module. I physically feel very good. Greetings to all the people of the nation and all the people of the world!" After that, Zhai Zhigang waved the red Chinese flag in space and moved along the orbital module by relying on the movable handrail outside the orbital module and the safety cables and hooks. He retrieved the solid lubricant experiment

▶ Diagram on Extravehicular Operation of Shenzhou 7 Astronaut

sample installed on the outside wall of the orbital module and handed it to Liu Boming. At 16:59, Zhai Zhigang returned to the orbital module and closed the hatch. When Shenzhou 7 flew around the Earth for the 31st time, it successfully released the miniaturized satellite.

Each small step of Zhai Zhigang in space had special historic meanings. It marked another major technological breakthrough from China in the manned spaceflight program and created a number of first-time achievements in China's manned spaceflights: a Chinese astronaut's debut spacewalk; the first time the national flag of the People's Republic of China had been flown in space; China's first self-developed feitian extravehicular space suit which had withstood the test of the space environment; the spacecraft and China's first data relay satellite Tianlian 1 conducted the first data relay experiment; the Shenzhou spacecraft carried a three-member crew into space for the first time; an astronaut collected a sample for the solid lubricant exposure experiment smoothly and carried out the first recoverable solid lubricant exposure experiment. All of this has laid a foundation for China's future construction of the space station, and made China the third nation in the world to have mastered the spacewalk technology.

Shenzhou 8 Mission: First Rendezvous and Docking in Space

In November 2011, the Shenzhou 8 spacecraft, as the chase vehicle, performed China's first in-space rendezvous and docking with Tiangong 1, the target vehicle that had been waiting a long time for this docking. Success of the Shenzhou 8 mission marked an important milestone in the second-step development strategy of China's manned spaceflight program showing that China had mastered automatic in-space

The Shenzhou Spacecraft:
Chinese Nation's Millennium Dream of Spaceflight Comes True

▶ Shenzhou 8

rendezvous and docking technology and set the stage for future manned docking and space lab construction.

Tiangong 1, launched on September 29th, 2011, was a new manned spacecraft that China had developed independently, and a target vehicle and simple space lab able to support rendezvous and docking for several times. Tiangong 1 weighs 8.6 tons, and measures 10.4 m in total height and 3.35 m in maximum diameter. It is made up of an experiment module and a resource module, with the former to accommodate facilities and supplies able to support 60 days' in-capsule work and the lives of the

▶ Shenzhou 8 Preparing for Rendezvous and Docking with Tiangong 1

China's Journey to Space: From Dream to Reality

▶ Tiangong 1 and LM-2F Complex on Launch Pad

▶ Tiangong 1

▶ Assembly of Tiangong 1 Target Vehicle

▶ Diagram on Work of Astronaut in Tiangong 1

The Shenzhou Spacecraft:
Chinese Nation's Millennium Dream of Spaceflight Comes True

▶ Diagram on Shenzhou 8's Entering into Target Search Stage

▶ Diagram on Shenzhou 8's Extending Docking Equipment for Docking

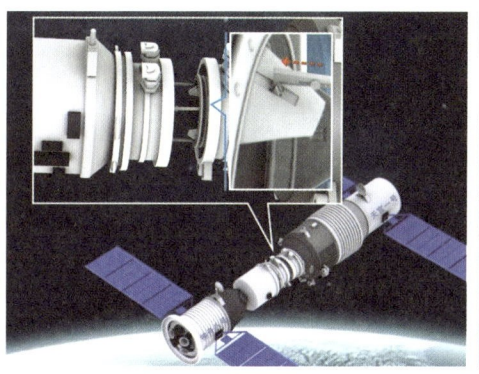

▶ Diagram on Shenzhou 8's Capture of Tiangong 1

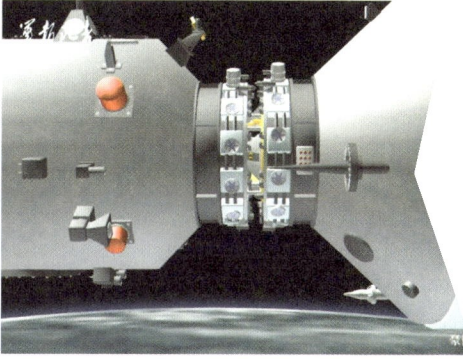

▶ Diagram on Rendezvous and Docking of Tiangong 1 with Shenzhou 8

▶ Diagram on Shenzhou 8's "Crash" into Tiangong 1

▶ Diagram on On-orbit Operation of the Tiangong 1 and Shenzhou 8 Complex

astronauts and the latter to provide power and energy.

After Shenzhou 8 took off into space, on its second day (November 2nd) when it was about 52 km away from Tiangong 1, it captured the signals of, and established communications links with, Tiangong 1. After that, it slowly approached Tiangong 1 and reduced its distance to 5000 m, 400 m, 140 m, 30 m... and carried out four "brakes" in the process. At 1:30 am on November 3rd, both Shenzhou 8 spacecraft and Tiangong 1 flew over Gansu and Shaanxi of China, and the two vehicles sensed through their sensing devices that they already contacted each other and then they performed a series of actions, including buffering, calibration, pulling closer and tighter and locking. Finally, the two vehicles were connected together which brought the first in-space rendezvous and docking to a successful end.

Shenzhou 8 and Tiangong 1 remained docked for 12 days and they carried out the second rendezvous and docking experiment on November 14th. On November 16th, Shenzhou 8 successfully separated with Tiangong 1 and returned to Earth safely.

Shenzhou 9 Mission: Travel of First Female Astronaut to Space

On June 16th, 2012, astronauts Jing Haipeng, Liu Wang and Liu Yang flew with the Shenzhou 9 spacecraft into space, embarking on China's new manned spaceflight journey. Shenzhou 9 mission bettered many records in China's manned spaceflight history: the first execution of space tasks by China's first female astronaut, the first implementation of astronaut manual docking, and the first "space visitors" to the Tiangong 1 target vehicle...

As there was a female astronaut on board, the Shenzhou 9 spacecraft was also equipped with an in-capsule clothing and accessory package, a set of pressure garments, toilet facilities and a sanitary bag for the exclusive use of female astronauts.

On June 18th, after docking of Shenzhou 9 with Tiangong 1, the three astronauts opened their binding belts, stood up from their seats and entered into the orbital module of the spacecraft, where they took off the space suits and changed into their blue work clothes. After a confirmation regarding the safety of Tiangong 1, astronaut Jing Haipeng opened the hatch of the experiment module of Tiangong 1 with a special "key" and then floated into Tiangong 1. Later, the other two astronauts also followed him into Tiangong 1, where the three astronauts replaced and installed relevant devices and equipment and conducted experiments.

Shenzhou 9 and Tiangong 1 remained docked for six days until June 24th when the two vehicles separated temporarily, so that astronaut Liu Wang could perform manual rendezvous and docking. Manual rendezvous and docking is a test of an astronaut's coordination between his eyes and hands, his carefulness during manoeuvering and

▶ Shenzhou 9 Spacecraft

China's Journey to Space: From Dream to Reality

psychological stability. To accomplish this, Liu Wang, responsible for this manual docking, was trained more than 1,500 times on the ground docking simulator. This helped him achieve technological proficiency and develop excellent psychological strength. Even if there were something wrong with the display system with no parameters for reference, he could operate docking precisely.

With the implementation of its Shenzhou 9 mission, China also created many first-time achievements in its manned spaceflight program: China conducted the first forward rendezvous and docking; a Chinese astronaut carried out his first manual rendezvous and docking; China first examined the manual control system of the spacecraft; Chinese astronauts paid their first visit to an on-orbit vehicle; China carried out the transportation and supply of crew and materials from ground to on-orbit vehicle for the first time; China assessed Tiangong 1's maximum capability in providing astronauts with working and living support for the first time; a female Chinese astronaut executed space tasks for the first time; Tiangong 1 performed tasks by using medium-term on-orbit residence support technology in China's manned spaceflight programs for the first time.

▶ Liu Yang in Training

The Shenzhou Spacecraft:
Chinese Nation's Millennium Dream of Spaceflight Comes True

▶ Shenzhou 9 Ready for Thermal Vacuum Test    ▶ Three-Module Docking of Shenzhou 9

Jing Haipeng
Liu Yang
Liu Wang

▶ On June 16th, 2012, LM-2F carrying Shenzhou 9 remote sensing satellite being successfully launched from Jiuquan Satellite Launch Center

▶ Shenzhou 9 Crew

China's Journey to Space: From Dream to Reality

▶ Diagram on Rendezvous of Tiangong 1 with Shenzhou 9

▶ Sectional View of Shenzhou 9 Spacecraft

Shenzhou 10 Mission: First Applied Spaceflight

On June 11th, 2013, LM-2F sent Shenzhou 10, with a three-person crew of Nie Haisheng, Zhang Xiaoguang and Wang Yaping, into space. The astronauts of Shenzhou 10 lived and worked in space for 15 days, making Shenzhou 10 China's longest human spaceflight mission to date, with the most tasks carried out, and marking China moving into the construction stage of the manned space station program. In this sense, Shenzhou 10 was a follower of previous spacecraft missions, but also a pioneer for future ones.

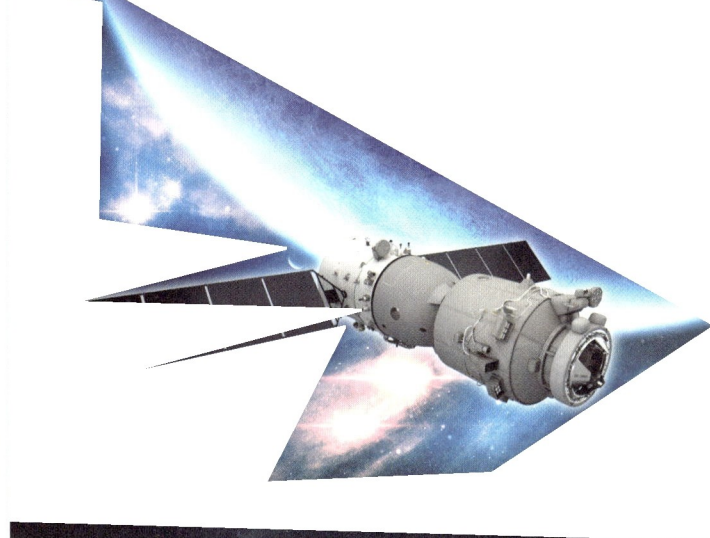

▶ Folding and Closing Fairing for Shenzhou 10

▶ Shenzhou 10

Instead of being a simple repetition of the Shenzhou 9 mission, the Shenzhou 10 mission was the first applied spaceflight mission of the Shenzhou manned spacecraft. It put into effect the transition from the development of spaceflight test spacecraft to the development of the

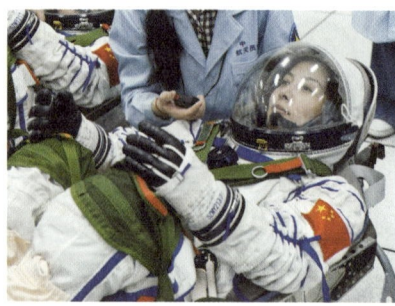

▶ Zhang Xiaoguang Engaged in Manual Rendezvous and Docking Training

▶ Stress Test for Pressure Garment of Wang Yaping

applied spacecraft, and took charge of the mission to provide on-orbit operation of Tiangong 1 with crew and material support and transport the crew and materials to and from space. Through this spaceflight mission, China further checked the performance and functionality of manned spacecraft and once again validated the reliability and safety of the increasingly established Shenzhou spacecraft. Meanwhile, China also examined the capabilities of the complex of Shenzhou 10 and Tiangong 1 in supporting the life, work and health of the astronauts, the functions, performance and coordination of all systems in performing the spaceflight mission and the abilities of the astronauts in carrying out spaceflight tasks. Moreover, the mission also included the research on the astronauts' space environment adaptability and space operation efficiency, the performance of space scientific experiments and on-orbit vehicle maintenance experiments, and the execution of the first space lecture by Chinese astronauts.

On June 13th, Shenzhou 10 entered into orbit and realized automatic rendezvous and docking with Tiangong 1 through remote ground guidance and automatic flight control. Following a series of preparations, the three astronauts entered into the orbital module of Shenzhou 10, where they took off their in-capsule space suits and changed into their blue work clothes. After that, mission commander Nie Haisheng and astronaut Zhang Xiaoguang first went into the docking channel and got ready to open the hatch of Tiangong 1. Standing at the hatch of Tiangong

1, Nie Haisheng looked relaxed and confident. He loosened the "key" to the hatch, that is, a handle of metal consistency, and exerted some force on the "key" making it whirl in space, a unique and wonderful experience for the astronauts in their space surroundings. After scientific researchers on the ground checked and confirmed the cabin environment of the Tiangong 1 target vehicle, the Beijing Aerospace Control Center gave the command to enter Tiangong 1 to the astronauts.

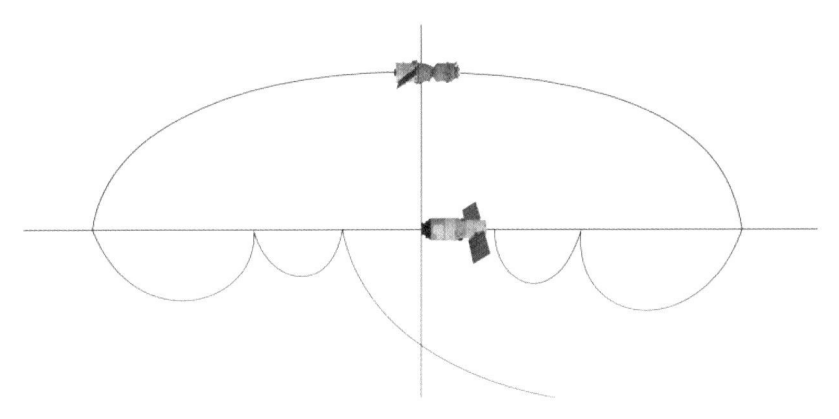

▶ Diagram on Flight of Shenzhou 10 around the Earth

On the afternoon of June 13th, Nie Haisheng gently opened the hatch of Tiangong 1 and the beige "floor" came into view. Then, Nie Haisheng waved his hands and "swam" into Tiangong 1 just like a fish. When it was shown on the large screen in the command hall of the Beijing Aerospace Control Center, the red Chinese knot in the module of Tiangong 1 appeared to be "greeting" the guests from its hometown. After getting a steady standpoint, Nie Haisheng turned around and beckoned Zhang Xiaoguang to follow him. Seeing the gesture of Nie Haisheng, Zhang Xiaoguang who had been waiting at the hatch also floated into Tiangong 1 with a confident smile on his face. The last crew member to enter into Tiangong 1 was the female astronaut Wang Yaping. Then, the three astronauts, arm in arm, faced the camera in the module and greeted ground scientific researchers with a wave of their hands, triggering warm applause in the command hall of the Beijing Aerospace Control Center.

To further improve the quality of space work and life of the astronauts, the technicians had: perfected garbage treatment in the module by increasing the variety and quantity of waste collection bags to facilitate the astronauts' sealing and storage of on-orbit domestic wastes, enriched space food by adopting individualistic food design for the astronauts, increased the kinds of food available onboard and enhanced the sensory acceptance of the food. The technicians also optimized the astronauts' work and life schedule by expanding time allowance for the working items.

Originally, the Tiangong 1 target vehicle only had hard floor in the section where the bicycle ergometer was installed. However, according to the astronauts on board Shenzhou 9, it would be more comfortable to walk on hard floor and also be easier to control posture. Therefore, Shenzhou 10 had the additional task of "floor replacement". On June 14th, after personnel on the ground checked and confirmed the cabin environment of the Tiangong 1 target vehicle,

▶ Shenzhou 10 and Tiangong 1 Ready for Docking

they instructed the astronauts to replace the interior cladding. Upon this instruction, the three astronauts in blue work clothes removed the original soft floor, paved the new hard cladding and installed new caging devices according to the working plan, using close coordination and collaboration. In addition, the astronauts also replaced the seal rings inside Tiangong 1. This was the first on-orbit maintenance job of Chinese astronauts. Plenty of experience and lessons from foreign manned spaceflight activities show that astronauts' mastery of various on-orbit vehicle maintenance technologies is of great importance to the progress of a long-term manned spaceflight program and to ensuring manned spaceflight safety. The on-orbit maintenance experiment during the Shenzhou 10 mission was valuable experience and laid the technological foundation for China's future construction of the space station.

On the morning of June 20th, the Shenzhou 10 astronauts gave their first space lecture to more than 60 million primary and middle school students nationwide from the Tiangong 1 target vehicle. The female astronaut Wang Yaping was the lecturer, the mission commander Nie Haisheng was her assistant and Zhang Xiaoguang was the cameraman. In the space lecture lasting about 40 minutes, the astronauts demonstrated special physical phenomena in the weightless environment in space, such as mass measurement, simple pendulum, gyro behavior and water film and water ball formation. This space lecture, as the first educational applied task in China's manned spaceflight program, manifested the idea of making the manned spaceflight program directly serve national education and further stimulated the young people's respect for science, love of space and enthusiasm for the exploration of the universe.

On June 23rd, with careful maneuvering by Nie Haisheng and working closely with Zhang Xiaoguang and Wang Yaping, the Tiangong 1 target vehicle and the Shenzhou 10 spacecraft carried out successful manual rendezvous and docking. After that, the astronauts Nie Haisheng, Zhang Xiaoguang and Wang Yaping visited the Tiangong 1 target vehicle once again to conduct relevant scientific experiments as scheduled.

On June 25th, the three astronauts left Tiangong 1 and returned to the re-entry module of Shenzhou spacecraft to get prepared for flying around the Earth. The end of the astronauts' visit to Tiangong marked the accomplishment of the Tiangong 1 target vehicle's historic missions. Since being in orbit on September 29th, 2011, Tiangong 1 had been in on-orbit operation for more than a year. During this period, Tiangong 1 received a total of six astronauts in two batches from Shenzhou 9 and Shenzhou 10 respectively, conducted many space science experiments and technological experiments and made significant achievements in scientific research.

After this, Tiangong 1 separated with Shenzhou 10, marking the success of China's first fly-around and rendezvous experiment.

After completion of all the preset tasks, the Shenzhou 10 spacecraft safely returned to Earth on June 26th and landed accurately at the predetermined zone. With the Shenzhou 10 spaceflight, China successfully accomplished the tasks for the first phase of the second step of China's manned spaceflight program, fully mastered vehicle rendezvous and docking technology under both automatic and manual conditions, manned cargo spacecraft development technology, space vehicle fly-around technology and space vehicle on-orbit maintenance technology, made breakthroughs in basic technologies for people-oriented design of the space area and for space station construction, and built the transportation system to and from space. All these show that the Shenzhou spacecraft is well-established as China's manned spaceflight and transportation system to and from space.

| The Shenzhou Spacecraft:
| Chinese Nation's Millennium Dream of Spaceflight Comes True

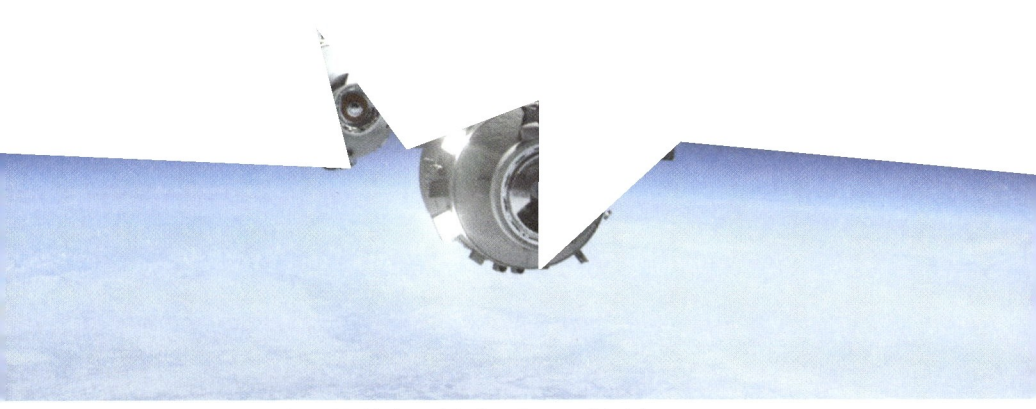

▶ Diagram on Separation of Orbital Module from Reentry Module

▶ On June 26th, 2003, the three astronauts Zhang Xiaoguang, Nie Haisheng and Wang Yaping (from Left to Right) Leaving the Module autonomously and Waving

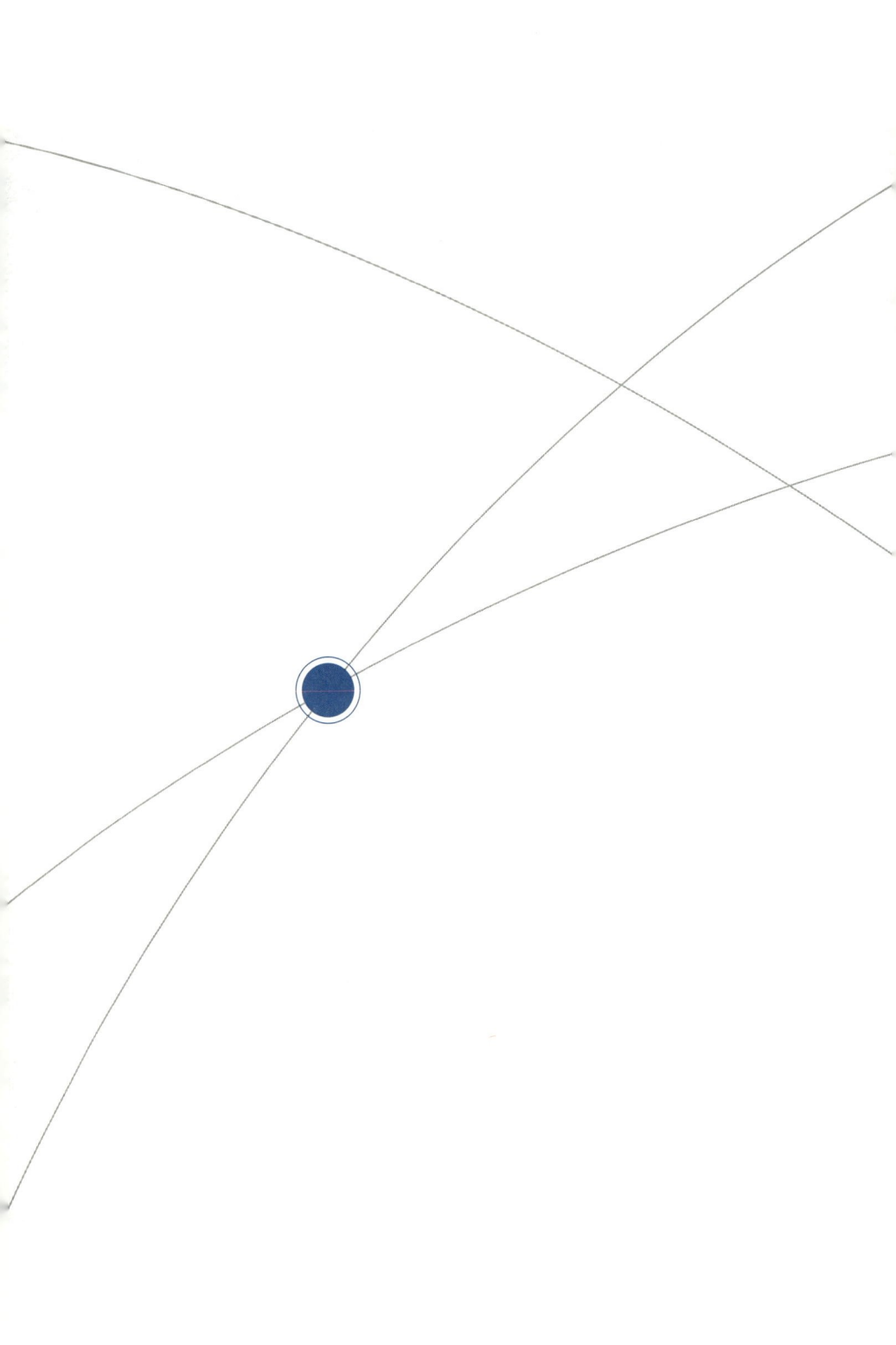

5

Travel of Chinese Astronauts to Space

The term "astronaut" is the general description for those people who operate manned spacecraft in space or work in space. In 2003, astronaut Yang Liwei flew around the Earth on board Shenzhou 5 for one day and became China's first spaceman. Since then, the term "astronaut" has resounded throughout China. To date, 10 Chinese astronauts have been to space in China's independently developed spacecraft.

Astronauts are the most recognisable symbol of manned spaceflights, but not everyone can qualify as an astronaut. Manned spaceflight activities are always risky and challenging and the quality of astronauts directly influences the success of the spaceflight missions and so astronauts must have solid dedication, a heightened capacity to learn and an extraordinary working ability, be psychologically stable and be in excellent physical condition. Saying that "astronauts are born to be astronauts" is never an understatement.

"Cradle" of Chinese Astronauts

In the northwestern outskirts of Beijing lies a bright pearl: the Beijing Aerospace Town. This beautiful and modern new town accommodates Chinese Astronaut Research and Training Center (CARTC), which is the Chinese astronaut selection and training base and also the only comprehensive institution dedicated to research on aerospace medicine, aerospace environment control and life-support technology and to the development of relevant products in China's manned spaceflight field. Chinese astronauts are born and raised here.

In 1968, with the concern and support of the older generation of scientists such as Qian Xuesen and Zhao Jiuzhang and to concentrate efforts on manned spaceflight development, CAS, the Chinese Academy of Medical Sciences (CAMS), the Academy of Military Medical Sciences and other relevant units jointly established the Aerospace Medicine and Engineering Institute, the predecessor of CARTC, to make preparations for aerospace medicine and engineering research and other relevant studies.

With more than four decades' development, CARTC has occupied three scientific research zones and two living quarters and constructed a great batch of ground simulation equipment and a complete, advanced lab system whose research findings and main technological capabilities have reached the world's top level. It has also become the third astronaut research and training center in the world after the Gagarin Cosmonaut Training Center in Russia and the Houston Space Center in the United States. Under the umbrella of the center are 13 professional labs, including the astronaut selection and training technology research lab, the space psychology lab, the aerospace medicine lab and the space nutrition and food engineering lab.

CARTC has selected and trained China's first batch of astronauts in Yang Liwei, Fei Junlong and Nie Haisheng and developed many products and devices, such as environmental control and life-support

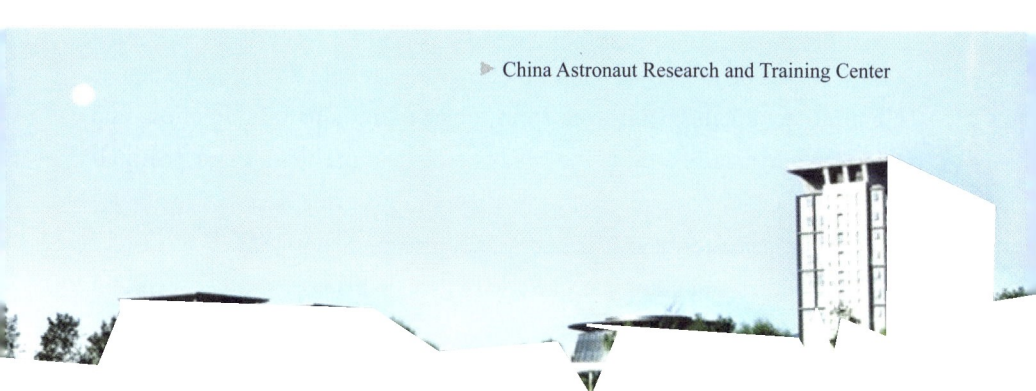

▷ China Astronaut Research and Training Center

equipment, space suits, space food, training simulators, onboard medical monitoring equipment and personal survival equipment for astronauts. In doing so it has made prominent contributions to the success of China's manned spaceflights.

Selection and Training of Chinese Astronauts

▶ Selection of "One in a Thousand"

Astronaut selection refers to the process of first screening astronaut candidates from a given population and then selecting crews qualified for manned spaceflights. Chinese male astronauts are chosen from active air force fighter pilots, while female astronauts come from active air force freighter pilots.

Requirements for pilots' participation in astronaut selection include: having a strong will, the spirit of self-sacrifice and dedication and superior psychological compatibility; being 160-172 cm in height and 55-70 kg in weight; being 25-35 years' old; having academic certificates at or above junior college level; having an accumulated flight experience of more than 600 hours, with excellent flight performance and no accidents; being free from tobacco and alcohol addiction; and having physical examination results from the previous three years all being of Class A standard. Physical conditions are a decisive factor for astronaut selection and it is necessary to reappraise the medical history of candidates and eliminate those with genetic disease or other recurrent diseases.

As astronauts have to face an extremely harsh environment, it is also necessary to examine the adaptability of candidates to the space environment. In the ascend phase of a spacecraft, astronauts have to withstand the noises and the feeling of being underneath a huge weight as a result of the rocket acceleration; in the on-orbit flight phase, they will face the difficult test of weightlessness and space radiation; in the re-entry phase, they will endure the feeling of being underneath a huge weight once again as well overheating; and upon the landing of the re-entry module, astronauts have to withstand the final test of landing impact and overload. Moreover, astronauts are also subject to a psychological compatibility assessment. In the confined space of a spacecraft, if astronauts fall into conflict, it will greatly affect the completion of a spaceflight mission.

In 1995, when China recruited the first batch of astronauts, 1,506 pilots met the basic conditions. However, about half of them were eliminated in the first round of detailed investigations, leaving only 886 pilots qualified for the original selection. After pilots passed the original selection, their family members were subject to general physical examination and medical history investigation. This round of elimination was more intense, as it washed out 90% of pilots who had passed into original selection, leaving 97 candidates. After careful evaluation by experts, a total of 60 pilots were told to attend the Air Force General Hospital, PLA, in Beijing, for a thorough examination. Finally, 14 astronauts were determined after more than 100 clinical medical examinations at the hospital.

Female Chinese astronauts are effectively required to meet the same requirements with regards physical quality, flight experience and state of mind. In addition, female astronauts should be married and, preferably, have already had a child. Spaceflight practices show that female astronauts can be just as qualified for spaceflights as male astronauts, and they even enjoy advantages over male astronauts both physiologically and psychologically. For example, female astronauts have greater resilience, a better tolerance of loneliness and a more stable psychological quality than male astronauts and also deal with some matters more sensitively, carefully and considerately than their male colleagues.

According to their different roles in a spaceflight mission, astronauts are further divided into commander, pilot, mission specialist, payload specialist, etc. For different kinds of astronauts, there are different selection standards and training requirements. For example, if we look at the Shenzhou 10 spaceflight mission, astronaut Nie Haisheng acted as commander.

▶ Astronauts Having Training in Simulation Module

▶ Tough, Specialized Training

A pilot must be trained for a long time before he or she becomes a real astronaut. This training generally lasts 2.5-4 years. A qualified astronaut

| Travel of Chinese Astronauts to Space |

- Nie Haisheng Entering Reentry Module for Training
- Female Astronaut Liu Yang Having Pulmonary Function Data Collected 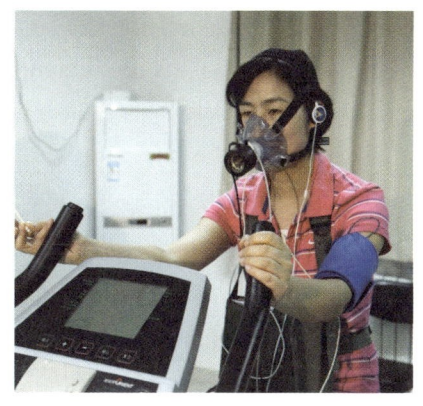
- Astronauts Jing Haipeng and Liu Wang Having Rendezvous and Docking Training in Reentry Module 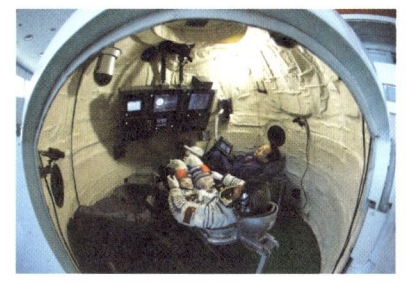
- Astronauts Having Lift-off Stage Operation Training in Spacecraft Simulator 

should master three skills: first, the skill to solve the five problems of being alive, i.e. to eat, drink, defecate, urinate and sleep in the weightless environment; second, the skill of operating and controlling a spacecraft, conducting experiments, Earth observation, telemetering and communications in the spacecraft and discovering and fixing problems in time; third, the skill to ask for help, save himself or herself and survive when he or she falls onto ground, into water or into any other harsh environment in case of abnormal re-entry.

Astronaut training can generally be classified into eight categories: physical fitness training, space environment adaptability training, psychological training, basic theory training, professional space technology training, flight procedure and task simulation training, life-saving and survival training and large-scale joint exercise. Each category of training also consists of different training programs and subjects. The training enables astronauts to gain the capability to carry out spaceflight missions from the perspectives of ideology, physical conditions, psychology, knowledge reserve and skills.

Astronauts should make themselves accustomed to special environmental factors in space and spacecraft and enhance their endurance through strenuous physical fitness training. The training of astronauts comprises 58 training subjects in eight major categories, which are called the 58 ladders to space by astronauts. With each ascent of a ladder, they make a step forward toward space.

To reduce or even avoid the space motion sickness of astronauts, researchers set rotational chair and swing, rotational bed and other training programs for astronauts. The rotational chair can carry out 360-degree rotations at a fast speed and can rotate up and down or back and forth at a maximum speed of 2.5 seconds/round. For the rotational chair training, the female astronaut Liu Yang ever said: "The rotational chair training in the air force only lasts for four minutes. However, the training for astronauts has to last for 15 minutes. After the fifth minute,

▶ Female Astronaut Liu Yang Examining Experiment Equipment in Tiangong 1
▶ Wang Yaping Making Training Preparations in Centrifuge

I thought that I had reached my limit, since I felt very uncomfortable. In spite of this, I could neither vomit nor ask them to stop. Although the instructors said that we could ask to stop once we could no longer endure it, no one has stopped midstream from the first batch of astronauts to us. As our body has a conditioned response memory for the rotational chair, if we vomit or stop during our first go, it will be more difficult to withstand it the next time. So, I tried my best to divert my attention by imagining that I was standing on a beautiful seaside. Since I got through my first go during training, I later became accustomed to it a little bit more each time."

Since astronauts will experience a huge acceleration during the ascent of a rocket, they need overweight endurance and adaptability training through a centrifuge. This is the most painful among all the training programs and subjects. When astronauts are engaged in the training, instructors observe and record facial expressions, conversations, various physiological indexes and response behaviors in the control room and thus make a scientific evaluation of the astronauts' overweight endurance and response characteristics. In this training program, when the centrifuge spins at a given speed, it can simulate the continuous overweight status during ascent and re-entry of a manned spacecraft. Normally, people can bear a load of three gravitational accelerations; however, astronauts have to withstand a maximum load of eight gravitational accelerations in their training, i.e. a weight equal to seven adults on their chest.

Professional space technology training is also a must for astronauts and trained astronauts can master various manipulative skills and relevant professional theoretical knowledge necessary for manned spaceflights. In particular, astronauts on board Shenzhou 7 and Shenzhou 9 spacecraft should also understand spacewalk technology and rendezvous and docking technology in order to complete their tasks smoothly. Therefore, before execution of the tasks, astronauts should be engaged in intense manual rendezvous and docking training on the ground. Liu Wang, China's first astronaut to perform manual rendezvous and docking in space, carried out simulated rendezvous and docking training on the ground more than 1,500 times and in the end, he managed to achieve a success rate of 100%. As a result, even when there are no parameters from the instrument for reference, he can operate precisely by means of visual display.

Unusual Space Life of Chinese Astronauts

In the weightless space environment, astronauts' life is completely different from that on the ground. Therefore, although it looks easy, eating, drinking, defecating, urinating and sleeping become difficult due to weightlessness.

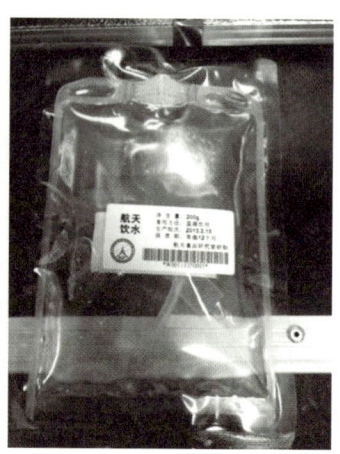

▶ Astronaut Having Meal in Space   ▶ Space Drinking Water

▶ Astronauts Being Particular about Eating and Drinking

Under weightless conditions, astronauts have to be particular about their meals. For a meal, each astronaut has a set of dishes, including a service plate, a spoon, a fork, a pair of safety scissors, etc. The service plate should be tied to a side of the astronaut's thigh, the spoon, the safety scissors and other dishware should be attached to the plate and the food should be fixed onto the plate with the Velcro of the plate. With these utensils, astronauts can easily take out chunks of meat and other solid food from open containers. However, when astronauts carry food to their mouth, they could put the food into their nose or eyes since their hands are not that flexible. It is better therefore, to put the food into the air, then approach the food to bite it. Additionally, astronauts should close their mouths while chewing their food; otherwise, the food debris will float everywhere and be difficult to remove. If the debris is inhaled and gets into the astronauts' lungs, it could cause serious problems.

In space, astronauts can never drink with cup because water does not flow downwards and so it cannot flow into mouth. So how do astronauts drink water? Generally, astronauts squeeze drinking water in a sealed bag which they take into their mouths through a straw. If it is a "solid drink", astronauts have to inject water into the sealed bag that contains the "solid drink" with an instrument similar to a "water pistol" and they can enjoy the drink after the "solid drink" has dissolved in water.

After having a meal, astronauts should place the food packing bags and the leftovers into waste collection bags and collect food debris with a food debris collector to avoid crumbs, water drops or food packages floating everywhere.

▶ Personal Hygiene of Astronauts in Space: Not Easy at All

Personal hygiene tasks necessary for people on the ground, such as brushing your teeth, washing your face, combing your hair, shaving, going to the toilet and having a shower, are not so easy in space as it turns out. Astronauts can brush their teeth in two ways: one is to chew a special gum used for oral cavity cleaning and the other is to wear a finger cot around your finger then brush your teeth with the finger and toothpaste before finally swallowing the toothpaste or spitting it out onto a paper towel. Astronauts cannot wash their faces with water as they do on the ground. Instead, they have to wipe their face with a wet towel dipped in a cleansing solution and then spread the towel onto a massage comb for hair combing.

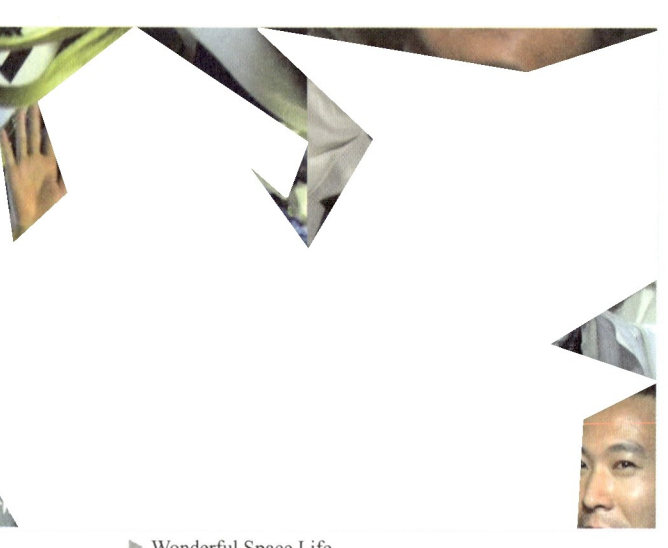

▶ Wonderful Space Life

For shaving, astronauts have to use a special shaving cream, which can absorb the astronaut's facial hair without the cream flying away. There is also the option of an electric razor, which directly collects the shaved beard.

So how do astronauts relieve themselves in space? In the Shenzhou 5 mission, the "toilet" Yang Liwei used was actually a urine collection device similar to a "disposable baby diaper", which can turn urine into a flocculent solid substance with water absorbing materials and perform deodorization. Nevertheless, this was not feasible for later multi-manned and multi-day spaceflight missions. When it came to the Shenzhou 6 mission, the orbital module was equipped with a "space toilet" developed by Chinese scientific researchers, which is a stool and urine collector. When astronauts want to relieve themselves, they need to bring the soft plastic hose of the toilet close to their excretory organ. After that, the air pump system in the toilet will suck the waste into a sealed container for stowage and at the same time, a deodorization device will work to remove peculiar smells. In fact, relieving themselves is also an important part of training for astronauts. The seemingly simple training is also divided into three stages: understanding theoretical knowledge, learning equipment operation and experiencing it in actual use; hence, it is not easy at all.

In space, astronauts cannot have a bath. They can only clean their body with a special paper towel and perform skin care with a tailor-made skin cream, but they can also change their underwear.

Since Tiangong 1's operations on orbit, astronauts have generally found the solutions to their personal hygiene problems in Tiangong 1. Female Chinese astronauts were also equipped with some cosmetics during their travels to space, including lipstick and cosmetic items equivalent to liquid foundation.

▶ The "Confused Sleep" of Astronauts

Many years of living on Earth has enabled humankind to develop both the daily routine of "going out to work at Sunrise and going back home at Sunset" and the biorhythm of a 24-hour cycle controlling the sleep-promoting hormones in human body. However, the astronauts in the Shenzhou spacecraft underwent 16 diurnal variations every 24 hours with one day lasting about 90 minutes for them. Therefore, they really suffered a "confused sleep". In the weightless environment in space, astronauts do not need a bed for sleep, as there is no difference between sleeping vertically and sleeping horizontally like back at home. Nevertheless, astronauts should attach themselves to the sleeping area; otherwise, after they fall asleep, their body will float about as if sleepwalking and may bump into instruments and equipment causing danger.

Generally, astronauts sleep for a shorter time in space and most astronauts suffer broken sleep. They will wake up after one or two hours' sleep and then continue to sleep for another one or two hours, or even more. After a long flight, astronauts are susceptible to mood fluctuations, deceptions, mirages and dreaminess, which can led to sleeplessness. For this reason, astronauts sometimes also need to take some soothing medicine or sleeping pills.

During the one-day flight of the Shenzhou 5 spacecraft in space, Yang Liwei had two naps lying on the seat. However, from the Shenzhou 6 mission onwards, as astronauts had to stay for a longer time in space, they needed enough sleep to ensure physical fitness and sufficient strength to guarantee smooth performance of the various space science experiments. As a result, later the Shenzhou spacecraft were also equipped with sleeping bags for astronauts, which were warm fabric bags with long zippers, attached to walls of the orbital module, with darker light and less noise. When astronauts want to sleep, they zip themselves into their sleeping bags and go to sleep. Afterwards, the sleeping bags resemble a huge silkworm larvae hanging from the wall.

After Shenzhou 9 spacecraft was docked with Tiangong 1, a separate sleeping area was especially designed for astronauts in order to minimize the impact of cabin noises and light on the sleep of the astronauts. In the sleeping area, there were rectangular sleeping bags and a household hanging sack attached to the wall, where astronauts could keep some small articles. The sleeping area was separated from other areas with an army-green, thick, movable curtain, which could prevent most noises coming from outside the area. Currently, the Shenzhou spacecraft has carried three astronauts to space each time and the sleeping area in Tiangong 1 could accommodate two astronauts at the same time with the third astronaut on duty. To welcome the arrival of female astronauts, the sleeping area also showed special attention to their physiological features and living needs. Moreover, in consideration of the fact that each astronaut has his or her own living habits, such as different tolerances towards light, technicians adopted a design for adjustable lighting, so that astronauts could adjust the brightness of lights whenever they needed according to their own preferences.

▶ Health of Astronauts in Space

In the special environment in space, situations injurious to the health of the astronauts may happen at any time. In view of this, telemetry information about ECG, blood pressure, breath, body temperature and other physiological parameters is transmitted to the ground control center, and medical monitoring doctors monitor changes of physiological indexes of astronauts in a real-time manner through ground medical monitoring facilities. In addition, medical monitoring doctors can also understand physiological and psychological situations of astronauts by directly talking with the astronauts while watching television images of astronauts. They can make comprehensive judgments of the health condition of the astronauts and carry out disease prediction and thus provide medical and psychological support to astronauts. When necessary, medical monitoring doctors can also promptly instruct astronauts to take pills.

For possible diseases to which astronauts may be susceptible in space, they have to carry more than one hundred drugs of various kinds. Pharmaceutical issues in space are more complicated and have rigorous and prudential requirements. For example, anti-nausea medicines on the ground generally have the adverse effect of hypersomnia, so for astronauts in space, medicines without such an adverse effect should be chosen and used.

Chinese aerospace medicine experts have also applied traditional Chinese medical science to spaceflight missions. Experimental results show that, traditional Chinese medicine has some effect on the prevention of space motion sickness and bone loss. Chinese astronauts took a decoction of herbal medicine before and after their spaceflight missions. After standard pharmaceutical and pharmacological and clinical experiments and approvals, this kind of traditional Chinese medicine was also brought to space for regular use by the astronauts.

▶ More Varied Leisure Time, Entertainment and Exercises in Space

Life in space does not only consist of work. When the spacecraft is flying around the Earth, astronauts also need a short break from their busy work with some entertainment available. If they are busy with work for days on end, they will feel strained. Therefore, when the ground personnel develop a spaceflight plan, they arrange the work of astronauts according to the working days on Earth so that the astronauts can have some time for relaxation, entertainment and exercises.

When China's first astronaut Yang Liwei was in space, the only recreation and entertainment for him might be to look out into space through the window, something that all astronauts like to do. Through the window, astronauts can see deep into outer space, they can see the Earth, the common homeland of humankind and they can see a Sunrise and Sunset appearing almost every forty minutes above the atmospheric layer of the Earth. This is a unique spectacle that Earth people will never see on Earth.

Before the Shenzhou 7 mission, the video connection between the astronauts and the ground personnel was unidirectional. In other words, astronauts could not see what was happening on the Earth. When it came to the Shenzhou 9 mission, the astronauts were connected to the ground via a two-way video, and they could see their family members when they were in space, which could help set their mind at ease. To make it possible for astronauts to have intimate conversations with their family members, technicians set up a special "private compartment" in Tiangong 1, where astronauts could not be "eavesdropped" no matter what they discuss with their spouses.

To inject more fun into the life of astronauts in space, there were some little surprises in Tiangong 1. The ground design personnel hid some small articles in Tiangong 1 and let astronauts find these articles themselves, in the hope of adding some fun to the monotonous space life of the astronauts. Some games that the astronauts played in space were not simply for the purpose of entertainment, but were related to scientific experiments. For the most part, astronauts were playing while doing scientific experiments, in order to examine changes of operant response and cognitive decisions in the weightless environment. These games can not only obtain scientific experiment data, but also make the life in space no longer boring.

In Tiangong 1, the astronauts could watch classical movies on a laptop and send emails with attachments as large as 8 MB, which was mainly achieved via the "Tianlian" data relay satellite in space. For example, if we look at the crew of the Shenzhou 10 spacecraft, Zhang Xiaoguang was always the one entertaining the other members of the crew so the commander, Nie Haisheng, gave the job of space entertainment to him. Zhang Xiaoguang prepared electronic books and melodious music for relaxation time for his two colleagues. In addition, Zhang Xiaoguang was also responsible for taking pictures and filming in space, for which he learned the professional techniques of "pushing, pulling, shaking and shifting". With a portable camera, he assisted Wang Yaping in accomplishing the space lecture, he recorded magnificent scenery of space and the life of his partners in space throughout the mission and he transmitted them to the ground in the form of high-resolution and high-quality images.

"Feitian" Space Suit: Special Space Suit for Chinese Astronauts

Space suits are garments astronauts should wear when in outer space. In the vacuum environment in space, nitrogen in human blood will change into its gas form and lead to expansion. Under these circumstances, if astronauts do not wear pressurized and gas-tight space suits or do not stay inside the pressurized and gas-tight re-entry module, they will put their lives in danger due to the great difference between body pressure and external pressure. At the same time, space suits should also supply breathable oxygen necessary for the human body, eliminate the carbon dioxide generated and control temperature and humidity in order to create an appropriate environment for survival of astronauts. Additionally, space suits can also protect astronauts from harm of space radiation and micrometeoroids.

A space suit consists of three layers: an external restraint layer, a middle gas tight layer and an internal heat removal layer. In terms of functions, space suits are classified into I.V.A. (Intra Vehicular Activity) and E.V.A. (Extra Vehicular Activity) spacesuits. I.V.A. spacesuits are relatively simple in structure and functionality and worn by astronauts during launch and re-entry of a spacecraft. They can protect the life safety of astronauts, and are generally tailor-made for each astronaut.

E.V.A. spacesuits are protective equipment, necessary for extravehicular activity, work done outside spacecraft. They can separate astronauts from the harsh environment of the space. It can be said that E.V.A. spacesuits are actually the smallest manned spacecrafts. China's first indigenous spacesuit is called Feitian. The domestically-made suit, costing 30 million yuan, is able to support four hours of extravehicular activity, and can be reused for five times. The suit, a combination of hard and soft, weighs 120kg, consisting of helmet, upper torso, lower torso, pressure gloves and boots, from top to bottom. With bands on limbs, the Feitian spacesuit is capable of adjusting the length of the upper arm, forearms and lower limbs, so that people with a height of 1.6m to 1.8m can wear it. It is worth mentioning that the upper limb joints, where biomimetic structures are cleverly applied, are able to move more easily. Besides, equipped with a small mirror in the wrists, astronauts can freely keep an eye on various switches on themselves.

The Feitian space suit backpack, measuring 1.3 m in height, has the out-capsule space suit life support equipment inside it and a hanging bag and a supplementary oxygen cylinder at its bottom. The helmet of a space suit is designed to have a broad field of vision and is equipped with cameras, which can shoot the extravehicular operations of astronauts. Both sides of a space suit are installed with a search light, which can light up the chest part of the space suit and facilitate operation of astronauts in a shadowy area. In addition, it also set up with alarm indicators in two sides, which will flash and give a spoken warning alarm in the case of any leakage of the garment.

Feitian space suit gloves contain an external fabric layer and two gas-tight layers, which are made of special isolated rubber materials and are able to withstand a temperature as high as 100 degrees Celsius. At the fingertip parts of the gloves, there is only one gas-tight layer to maintain the sense of touch, and the palm parts of the gloves are furnished with convex granular rubber for anti-skid purposes, which enables astronauts to hold articles with a diameter as small as 25 mm, such as pencils.

Chinese Space Food

Space food has the highest requirements out of any kind of food. While ensuring absolute food safety, space food and their packages should also stand the test of various space environmental factors, such as impact, vibration and acceleration. Moreover, space food should also be light in weight, small in volume, rich in nutrition, easy to digest and free from food debris.

The structure and functions of space food depends on the duration of a spaceflight mission, the restrictive conditions of a spacecraft and the complexity of the tasks to be undertaken. Generally, they are required to be geared to the special environment in space and the engineering design of a spacecraft, and meet physiological and psychological needs of astronauts to the maximum extent. This is to provide sound nutrition and food safety and ensure the success of a spaceflight mission. From the perspective of nutriology, nutrients in space food should be properly adjusted according to physiological change indexes of astronauts. For example, space food should supply sufficient quality protein for muscular atrophy and enough calcium as well as a proper calcium-to-phosphorous ratio and Vitamin D for loss of bone calcium.

During the Shenzhou 5 mission, astronaut Yang Liwei only stayed in space for 21 hours. Therefore, the space food on board was instant food with no need for heating and adding water. When it came to the later "multi-manned and multi-day" mission, the astronauts had a wider living space with a special "dining area". They also had three meals every day in space just like on the ground, with their food rich in variety and with distinct Chinese characteristics.

Space food must suit the tastes of the astronauts. Studies have shown that, while in space, people are not as sensitive to sour, saline and other tastes. That explains why astronaut Nie Haisheng said he was very fond of spicy, garlic-tasting food when in space. After returning to Earth, astronaut Fei Junlong also listed foods he thought were delicious: shrimp meat, braised beef with brown sauce, steamed gluten and Cantonese-style grilled pork.

In 2012, the astronauts on board Shenzhou 9 spent the traditional Chinese Dragon Boat Festival in space where they had "eight-treasure rice pudding" in place of the traditional sticky rice wrapped in bamboo leaves for the festival. The Shenzhou 10 mission also coincided with the Dragon Boat Festival, and the astronauts on board had the traditional sticky rice stuffed with red bean paste and wrapped in bamboo leaves exquisitely prepared by scientific researchers. Different from the ones on the ground, the sticky rice adopted a flat shape for the convenience of storage and heating. To enable the astronauts to feel as if they were celebrating the festival at home, the scientific researchers also wrapped each piece of steamed glutinous rice with a layer of bamboo leaves.

Astronauts could also have fruit in space, such as strawberries, apples, bananas, pineapples, honey peaches and Hami melons that had undergone low temperature freezing, water removal and other procedures. Although they are "dried" fruits, they are still bright in color and delicious. Most astronauts like these "dried" fruits. In addition, since Chinese people like drinking tea, green tea, black tea and other drinks were also prepared for astronauts. However, these drinks were not in liquid form. Instead, they were solid drinks in the shape of a small brick. When astronauts wanted to drink them, they had to perform some blending before they could enjoy the satisfying taste.

With such delicious space food being rich in nutrition, it not only ensures the astronauts have sufficient physical strength for performance of the spaceflight mission, but has also become a form of space entertainment, don't you think?

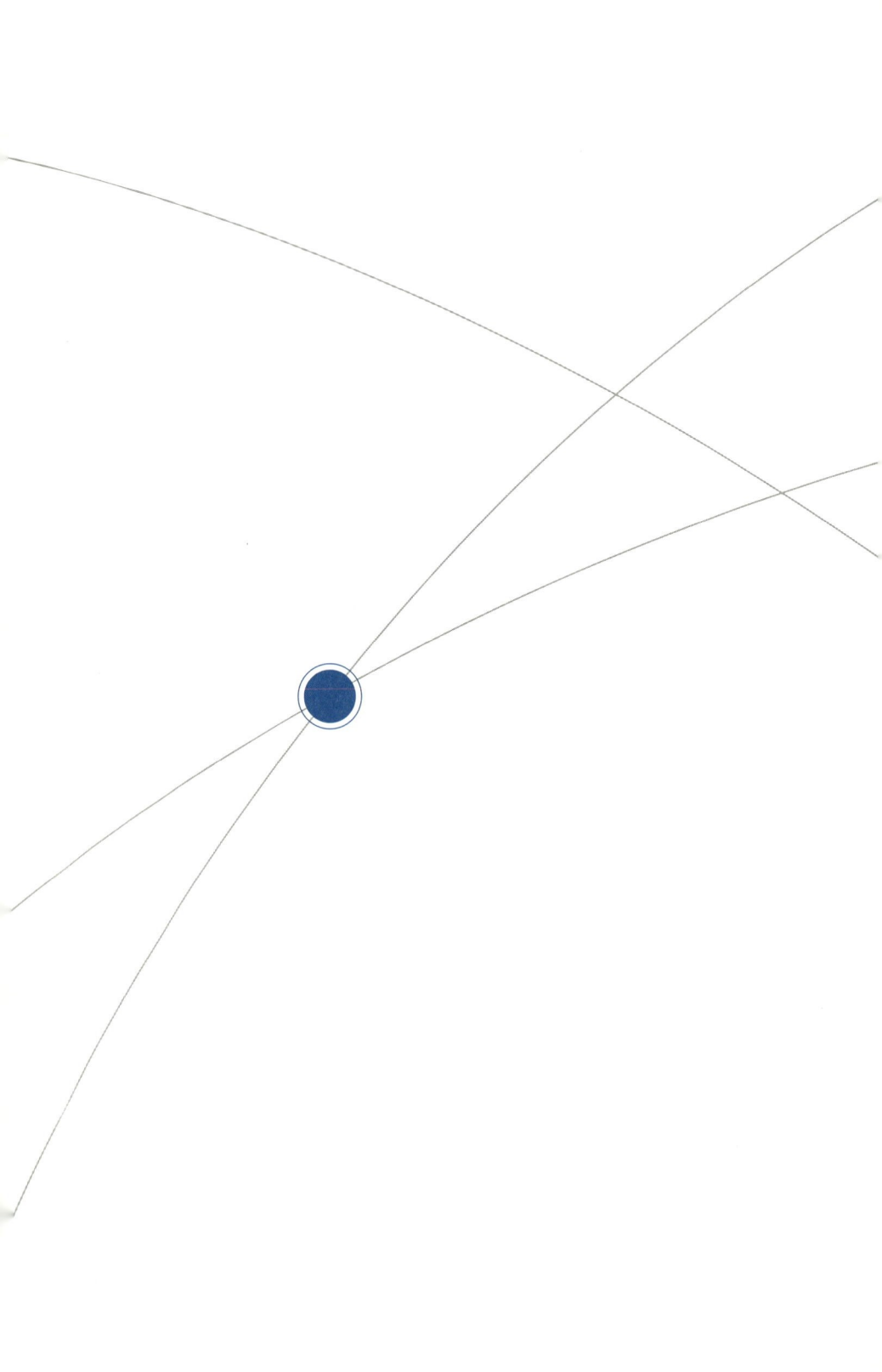

6

The Wonderful Beginnings and Progress of the Chang'e Program

On October 24th, 2007, China's first lunar exploration satellite, Chang'e 1, was successfully launched atop an LM-3A carrier rocket from the Xichang Satellite Launch Center, marking the beginning of China's deep space exploitation after more than five decades' development of the aerospace sector.

From the organization of relevant experts for preliminary analyses and the demonstration of the necessity and feasibility of lunar exploration in 1994 up to official approval and establishment of the lunar exploration program in January 2004, over 50 academicians and more than 150 experts were engaged in the program demonstration and the preparations to carry out the exploration. They completed a lot of preliminary analyses, demonstrations and conceptual design work, laying a solid foundation for the accomplishment of the plans for the lunar exploration program on schedule.

From 2004 to 2012, scientists and technicians working on the lunar exploration program broke new ground, overcame a series of technological difficulties and developed and launched two lunar exploration satellites, namely Chang'e 1 and Chang'e 2, bringing a successful end to "orbiting", the first phase of the lunar exploration program and setting the stage for "landing", the second phase of the program.

First Lunar Exploration by Chang'e 1

The Chang'e 1 satellite was China's first lunar orbiting satellite and also China's first deep space probe to overcome the Earth's gravitational pull to visit a celestial body outside the Earth, marking China's first step toward deep space exploration.

▶ Missions and Objectives of Lunar Orbiting Exploration

With the Change'1 satellite, China generally mastered the basic lunar exploration technologies, carried out the first scientific exploration of the Moon and formed the initial lunar exploration and aerospace engineering system, accumulating experience for subsequent lunar exploration work.

Lunar orbiting served the first phase of China's lunar and deep space exploration program. Based on the analyses of the international lunar exploration history, the current situation and the development trends and according to China's needs for scientific and technological development and actual capabilities, scientists and technicians put forward four scientific objectives for Chang'e 1: first, to acquire three-dimensional lunar images; second, to probe contents of lunar elements, substance types and their distribution; third, to explore lunar soil

▶ Chang'e 1 Satellite

features; and fourth, to exploit the space environment between the Earth and the Moon.

China is the first country in the world to take 3D photos of the entire lunar surface, which has great scientific implications for research about the terrains and geological structures of the lunar surface and the lunar craters. Study on the lunar landscape features can also provide the most direct evidence for lunar evolution history. Making use of existing exploration findings from foreign countries and drawing experience and lessons from foreign lunar exploration activities, China selected optimum exploration targets, optimized its technological implementation approaches and did something others had never done before in a progressive and innovative manner.

▶ Appearance and Configuration of Chang'e 1

The Chang'e 1 satellite adopted a hexahedron shape with a length of 2.22 m, a width of 1.72 m and a height of 2.2 m and had both of its sides installed with a large extendable solar wing with a span able to reach 18 m upon full spread of the wing. The satellite weighed 2,350 kg, operated in the circular polar lunar orbit at an altitude of 200 km and had a service life of one year.

The Chang'e 1 satellite was developed on the DFH-3 satellite platform. The so-called satellite platform with scientific instruments resembles a truck with various kinds of cargo. The satellite platform provides these scientific instruments with installation positions, energy supply, attitude guarantee, temperature environment, data management and other conditions, while the scientific instruments are responsible for the completion of lunar exploration tasks.

The Chang'e 1 satellite consists of the structure subsystem, the guidance, navigation and control subsystem, the power supply subsystem, the thermal control subsystem, the monitoring control and data transmission subsystem, the data management subsystem, the propulsion subsystem and the scientific exploration instrument subsystem, each of which not only perform their own functions to ensure normal operation of the satellite in space, but also carry out scientific exploration and investigation.

The lunar surface is composed of different elements and mineral substances, which are shown by distinctive characteristic spectra or rays in specific radiation bands. Therefore, the instruments on the Chang'e 1 satellite probed the visible lights near infrared spectra, gamma rays and x-rays emitted by substances on the lunar surface. The features of different wave bands or rays can tell the amount and distribution of various elements and minerals on the lunar surface.

The microwave radiometer on board the Chang'e 1 satellite is an instrument to assess the depth of the lunar soil. Microwave remote sensing has around-the-clock observation capability and a resolution

▶ Chang'e 1 Satellite Ready for Docking with Rocket

of dozens of centimeters. As a result, it was one of the high-tech areas that enjoyed the greatest priority in various countries around the world. Since the Moon is far away from the Earth and mankind does not have a thorough knowledge of the surrounding space environment of the Moon and the lunar surface, lunar exploration by means of microwave remote sensing remains very difficult.

The Chang'e 1 satellite exploited the space environment between the Earth and the Moon by using the high-energy solar particle detector and the solar wind ion detector.

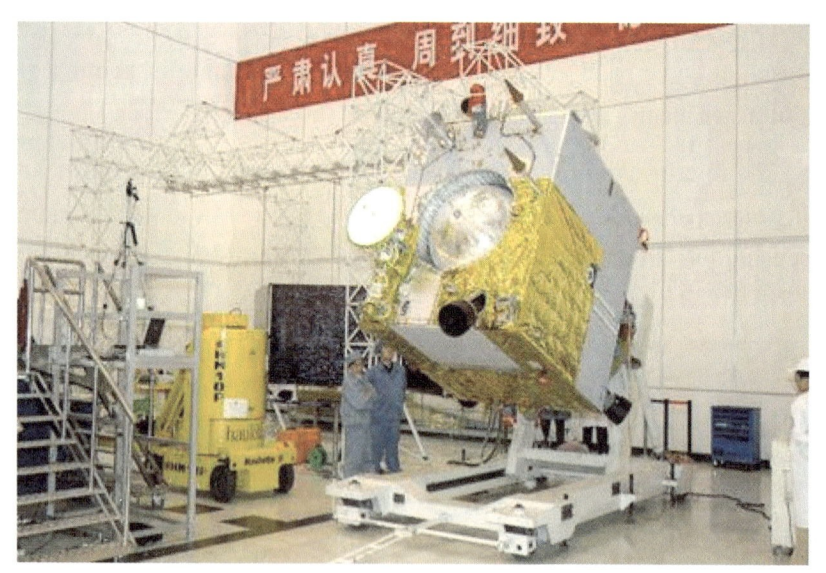

▶ Assembly of Chang'e 1 Satellite

To satisfy the aspiration of all Chinese people to participate in the lunar exploration program, COSTIND decided to carry 30 pieces of music representing the best parts of Chinese culture on board the Chang'e 1 satellite. After online voting by citizens and an expert review, 30 pieces of classical music were selected, including "My Motherland", "Ode to the Yellow River", "In the Place Far, Far Away", "The Story of

Spring", "Songs of the Seven Sons", "Girl from Ali Mountain", "High Mountains and Flowing Streams", etc. In November 2007, Chang'e 1 entered into the lunar orbit and transmitted back the melodious music from space, 380,000 km away from the Earth, and people heard the music from space through television and radio.

▶ Five Major Technological Breakthroughs in Lunar Exploration

First, they had to solve the flight track design problem of Chang'e 1. For satellites China had launched before Chang'e 1, their farthest distance to the Earth had not exceeded 70,000 km. However, Chang'e 1 had to travel into the vicinity of the Moon, 380,000 km away from the Earth. For this, it had to overcome the gravitational pull of the Earth and move into the gravitational pull of the Moon before finally finding a position next to the Moon in which they can move around the Earth together. It must neither crash into the Moon nor pass close by the Moon. Instead, it should "brake" accurately at the correct time when it approached the Moon and then enter into the lunar orbit at a reduced speed. In other words, the whole flight of Chang'e 1 had to go through the following stages: the phasing orbit, the Earth-Moon transfer orbit, the lunar capture orbit and the lunar orbit. The Chang'e 1 satellite is also the first lunar exploration satellite in the world to adopt such a flight track.

Second, they had to carry out long-distance monitoring, control and communications with the Chang'e 1 satellite. When the signals given by the Chang'e 1 satellite reached the Earth, their intensity was lower than one-hundredth that of the signals sent by a geostationary orbit satellite. Thus, there arose the question how to make the ground station receive satellite signals accurately, meanwhile enabling the satellite to receive the instructions from the ground? In consideration of the current state of China's aerospace surveillance control network and to solve the surveillance control problem of the Chang'e 1 satellite, scientists and technicians proposed a combination of the aerospace monitoring control network and the astronomical observation network of CAS for orbit determination, so as to improve the orbit determination precision of the Chang'e 1 satellite and meet the needs of lunar exploration.

To further enhance the measurement and control coverage and reliability of the Chang'e 1 satellite, China also cooperated with ESA, Australia, Spain and Chile and worked together with the ground measuring stations in Kourou, Nova Scotia, Las Palmas, San Diego, etc., to develop a global station deployment pattern combining Chinese

▶ Flight Trajectory Diagram of Chang'e 1 Satellite

space tracking ships and ground stations and foreign measurement and control stations. Through international cooperation and global station deployment, the Chang'e 1 satellite achieved a measurement and control coverage up to 98%.

Third, they had to overcome the difficulty of satellite orientation among the Sun, the Earth and the Moon. When the Chang'e 1 satellite flew into space, it needed to choose some celestial body as its reference point in order to determine its orientation in space. Previous Earth-orbiting satellites all took the Earth as their reference points, and could determine their attitude against the Earth with the infrared sensitive device suitable for the Earth environment on board. However, since there is no atmospheric layer around the Moon, there is thus no stable infrared radiation belt, which excludes the possibility of using an infrared sensitive device as the sensor for the Moon. Therefore, scientists and technicians designed the ultraviolet sensor suitable for lunar remote sensing and made it the "eye" of the satellite, with which the satellite could aim it at the Moon, obtain various lunar data with scientific instruments on board and complete the investigation tasks.

While aiming at the Moon, the Chang'e 1 satellite needed to also aim at the Sun and the Earth, as the solar wing of the satellite needed to keep adjusting its direction so as to be aligned to the Sun and obtain energy. The directional antennae of the satellite must be pointed to the Earth in order to establish the wireless communication link for data transmission and measurement and control signals between the satellite and the Earth and send data to the ground and receive instructions.

Fourth, they had to ensure that the Chang'e 1 satellite could withstand the unprecedented limit temperature difference. When the Chang'e 1 satellite flew around the Moon, it also had to encircle the Earth together with the Moon and even had to revolve around the Sun with the Earth and the Moon. Therefore, the satellite was faced with an extremely complicated and harsh thermal environment affected by the Sun, the Moon, the Moon's shadow, lunar eclipse and space coldness. This required the Chang'e 1 satellite to have a highly efficient and reliable thermal control system to ensure that all exploration instruments on board were in their appropriate working temperature range. At the same time, scientists and technicians also helped the Chang'e 1 satellite wear a special "coat", a surface heat insulation layer. It could preserve heat when the satellite was in an environment with an extremely low temperature and assist the satellite in heat radiation when the internal temperature of the satellite was too high.

Finally, they had to guarantee long-term around-the-Moon flight of the Chang'e 1 satellite. To enable the exploration instruments on board the Chang'e 1 satellite to obtain measurement data on most areas of the lunar surface as quickly as possible, with the same resolution, scientists and technicians chose the circular polar lunar orbit for the satellite. To ensure that the Chang'e 1 satellite would complete its one-year exploration tasks smoothly, calculations and analyses show that the ground control personnel had to adjust the satellite operation orbit about every 50 days so as to maintain the normal circumlunar trajectory.

▶ Journey to the Moon with Twists and Turns

The Moon is 380,000 km away from the Earth and also revolves round the Earth. As the Chang'e 1 satellite flew from the Earth to the Moon, it had not only to cover the distance of 380,000 km but it also had to go through the change from effect of the gravitational pull from the Earth to the effect of the gravitational pull from the Moon. To ensure the Chang'e 1 satellite entered the Moon's vicinity at the proper time and place, it was necessary to design the flight track in advance.

The flight track of Chang'e 1 satellite consisted of four stages: the phasing orbit, the Earth-Moon transfer orbit, the lunar capture orbit and the lunar orbit. In the four stages, the time and the motion state of the satellite upon its entry into the Earth-Moon transfer orbit, especially its position, speed and direction, were very critical. If the timing was incorrect, the Chang'e 1 satellite would fail to find the Moon and have no way to accomplish its mission.

On October 24th, 2007, the Chang'e 1 satellite was successfully launched on the top of an LM-3A carrier rocket. The satellite separated

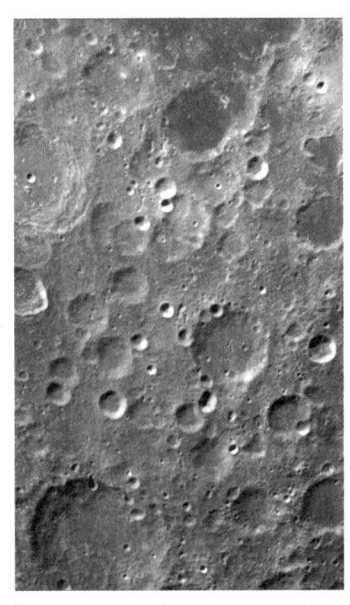

| The Wonderful Beginnings and Progress of the Chang'e Program |

▶ Local 3D Landscape Map on the First Lunar Image Released by NASA and Transmitted back by Chang'e 1 Satellite

▶ First Lunar Surface Image Transmitted Back by Chang'e 1 Satellite

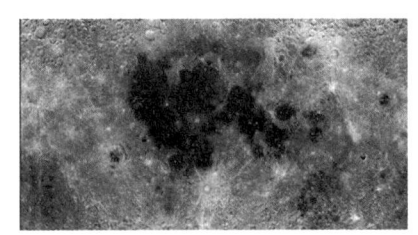

▶ First Image of the Entire Moon

with the rocket 24 minutes after the rocket fired and entered the super geosynchronous transfer orbit at a perigee altitude of 205 km and an apogee altitude of 50,900 km and with a cycle of about 16 hours. According to the flight program, the Chang'e 1 satellite extended its solar wing and then its directional antennae, changed into the cruise flight attitude, finished all actions in the launch stage and entered into the phasing orbit stage.

On October 25th, the Chang'e 1 satellite reached the apogee after flying on the super geosynchronous orbit for one and a half cycles. To enlarge the coverage of ground measurement and control, the ground personnel quickened the satellite at the apogee for the first time to raise its perigee altitude from 205 km to 593 km.

On October 26th, Chang'e 1 satellite was successfully accelerated for the second time, and entered into elliptic orbit at a perigee altitude of 593 km, an apogee altitude of 71,600 km and a cycle of 24 hours. The satellite travelled on the elliptic orbit for three days and on October 29th, with the third acceleration, the satellite moved into the large elliptic orbit at an apogee altitude of 119,800 km and a cycle of 48 hours.

On October 31st, after the Chang'e 1 satellite accomplished flight in the phasing orbit stage, its fourth acceleration increased the apogee altitude to 405,000 km. With the accelerated speed, the satellite went into the Earth-Moon transfer orbit. Until then, the Chang'e 1 satellite had flown for seven days and gone around the Earth seven times, during which it underwent acceleration at the apogee once, at the perigee three times and raised its speed to 10.58 km/s. When the satellites began to speed up, some of the scientific instruments on board began to work.

After the Chang'e 1 satellite was on the Earth-Moon transfer orbit, it flew on the orbit at a perigee altitude of 600 km and an apogee altitude of 405,000 km for five days and covered 436,600 km in order to slowly get closer to the Moon. During this process, the ground personnel successfully performed a mid-course correction of the orbit for Chang'e 1 satellite on November 2nd. Since this mid-course correction was precise and accurate and had achieved the desired objectives, the scheduled second correction turned out to be unnecessary.

On November 5th, the Chang'e 1 satellite flew over the Moon. Then, the satellite had to reach the perilune at the predetermined time and speed and "brake" precisely to meet with the Moon; otherwise, the Chang'e 1 satellite would miss its opportunity. In this sense, the "braking" was crucial during the entire journey to the Moon and was also the key to success of the mission. At 11:15 on November 5th, the Chang'e 1 satellite braked for the first time at the perilune, which was then captured by the gravitational pull of the Moon and brought into the elliptic polar lunar orbit at a perilune altitude of 212 km, an apolune altitude of 8,617 km and a cycle of 12 hours. Since then, Chang'e 1 has served as a circumlunar satellite.

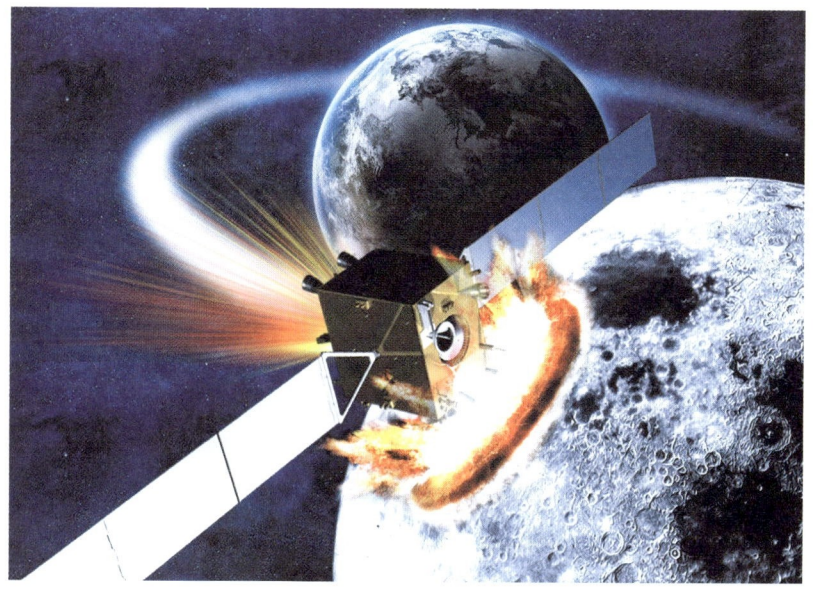

| The Wonderful Beginnings and Progress of the Chang'e Program |

▶ Chang'e 1 Satellite's Crash into the Moon

In the subsequent two days, the Chang'e 1 satellite also carried out braking twice at the perilune. On November 7th, the satellite finally reached its destination, the circumlunar working orbit, after having covered 1.8 million km.

▶ **Final Home of Chang'e 1 Satellite**

On November 7th, 2007, the Chang'e 1 satellite entered into the circular polar lunar orbit with a cycle of 127 minutes and at an altitude of 200 km, bringing the longest "march" in China's aerospace history to a perfect end. The Chang'e 1 mission carried out "correct launch, accurate injection, precise measurement and control, perfect orbital transfer and successful lunar orbiting operation".

On the morning of November 26th, 2007, CNSA officially released the first lunar surface image transmitted back by the Chang'e 1 satellite. Release of the first lunar image marked a great success of China's first lunar exploration mission. At the same time, the voice and the music from "Sing a Song of Praise to the Motherland" were also transmitted from the satellite back to the Earth on November 26th. After that, CNSA also released the images of the far side of the Moon, 3D images of the lunar surface and the images of the polar regions of the Moon. According to images and data obtained by the Chang'e 1 satellite, Chinese scientists found the highest peak on the Moon, which is 9,840 m high, almost 1,000 m higher than the Mount Everest.

The second phase of China's lunar exploration program is aimed at soft landing and has a big technological scope. To reduce risks of subsequent lunar exploration missions and accumulate experience controlling the collision with the Moon and orbit determination, it was decided that Chang'e 1 would hit the Moon at the end of its service life. Therefore, on March 1st, 2009, with precise control of the scientists and technicians at the Beijing Aerospace Control Center, the Chang'e 1 satellite underwent deceleration, falling then crashing... During the crash, the CCD on board the Chang'e 1 satellite still worked, transmitting back clear real-time images. Until then, after 494 days' flight, the tranquil and remote Moon finally became the final resting place of China's first "envoy to the Moon". With precise implementation of the "controlled collision", the first phase of China's lunar exploration program came to a perfect end.

Second Step in Lunar Exploration by Chang'e 2

On October 1st, 2010, the Chang'e 2 lunar exploration satellite was launched on the LM-3C carrier rocket from the Xichang Satellite Launch Center. After five days' travel, the satellite met with the Moon on October 6th and went into lunar orbit to perform technological experiments and scientific exploration tasks.

▶ Born for "Landing" after "Orbiting"

Soon after official establishment of the Chang'e 1 mission, the state decided to introduce a backup satellite for the first phase of the lunar exploration program to reduce the risks of the lunar orbiting exploration program and guarantee complete success. After the successful lunar orbiting exploration program, in December 2007, the leading group of the lunar exploration program organized demonstrations of applicable programs about the backup satellite. The newly-founded State Administration of Science, Technology and Industry for National Defense (SASTIND) continued to coordinate experts to hold further demonstrations. Since the second phase of the lunar exploration program will involve many key technological breakthroughs with a large technological scope and great difficulty in its implementation, it was determined after repeated demonstrations that the Chang'e 2 satellite would be sent into space as a pilot technology satellite before launch of China's first lunar lander, Chang'e 3, so as to reduce risks during the second phase of the program. The Chang'e 2 satellite would examine and validate some key technologies for the realization of soft landing by Chang'e 3 and Chang'e 4, accumulate relevant experience and carry out scientific lunar exploration and research based on the Chang'e 1 mission.

▶ Innovative Breakthroughs Based on Chang'e 1

Based on the successful research and development experience foundations of Chang'e 1, Chang'e 2 "travelled faster" than Chang'e 1 satellite (it only took 5 days to reach the destination), "approached closer" (it operated in 100-km orbit), and "saw more clearly" (it could choose the right time to descend into a lower orbit for accurate imaging of specific regions of the Moon).

As the "trailblazer" for the second phase of the lunar exploration program, the Chang'e 2 satellite shouldered the validation and exploration mission. More precisely, it was designed to validate the six key technologies, namely, the direct injection technology, the lunar capture technology, the x-band measurement and control technology,

▶ Chang'e 2 Lunar Exploration Satellite Successfully Launched atop LM-3C Carrier Rocket

| The Wonderful Beginnings and Progress of the Chang'e Program |

the orbit manoeuvre technology, the data transmission technology and the high-resolution imaging technology. At the same time, it needed to conduct more comprehensive exploration of the Moon from a lunar orbit 100 km away from the Moon. It especially required precise exploration of the alternative landing zone of Chang'e 3, so as to pave the way for safe landing of Chang'e 3. Chang'e 2 made successful breakthroughs in these aspects by means of independent innovation.

(1) The carrier rocket sent Chang'e 2 directly into the Earth-Moon transfer orbit, therefore flying, in a faster way, straight to the Moon. The key to solve this problem was the design of the Moon-Earth transfer orbit. The LM-3C rocket used for launching the Chang'e 2 satellite was derived from LM-3A with two boosters added, which increased the flight speed of the Chang'e 2 satellite to 10.9 km/s directly. As a result,

Chang'e 2 was able to skip the phasing orbit, directly enter into the Earth-Moon transfer orbit and shorten the travel time to the Moon by seven days.

In addition, Chang'e 2 adopted a "zero window" launch, which saved 180 kg of fuel when compared with the back edge launch. This launch mode not only shortened the spaceflight time, but also made full use of the capacity of the carrier rocket to reduce fuel burn, thus enabling the satellite to execute more experimental tasks and also delivering a longer service life.

▶ Chang'e 2 Lift into Thermal Vacuum Tank for Test

(2) The lunar orbit 100 km away from the Moon was the "mission orbit" of the Chang'e 2 satellite, and to validate and realize the orbit capture technology was also the key to the success of the Chang'e 2 mission. This orbit has two prominent features: first, it was unprecedentedly close to the Moon, as the orbital altitude of Chang'e 2 satellite was only 100 km, but Chang'e 1 operated on the 200 km orbit; second, the orbital cycle was shorter than ever before, for it only took

Chang'e 2 about 118 minutes to fly around the Moon once after braking three times.

After the Chang'e 2 satellite made adjustments and operated for some time in the 100 km circular lunar orbit, it lowered by itself at the far side of the Moon to an elliptic orbit with a perilune altitude of 15 km and an apolune altitude of 100 km, and then flew to the Sinus Iridum on the near side of the Moon for clear imaging with a resolution higher than 1.5 m. This orbit lowering and exploration was the biggest test for the Chang'e 2 satellite, and by seizing the right opportunity to undertake orbit lowering, the satellite found the solution to passing this test. Since the orbit lowering opportunity determined the subsequent position of the perilune, if the right opportunity was missed, a deviation would occur between the 15 km perilune position and the alternative landing zone. Practices show that the Chang'e 2 satellite not only seized the opportunity but also landed at the alternative position accurately.

(3) The Chang'e 2 satellite also achieved the breakthrough of precise exploration and obtained the clearest lunar surface images with a resolution in excess of 10 m and a resolution higher than 1.5 m at the 15 km perilune. Therefore, the Chang'e 2 satellite was also equipped with a landing camera and 3 monitoring cameras. Among them, the landing camera would serve as an "eye" for the successor, Chang'e 3, to search for a safe landing site during its soft landing. It could take photos of the preselected landing zone in a real-time manner and also shoot the landing process of Chang'e 3, so that the Chang'e 3 satellite could avoid an inappropriate landing site by itself based on the photos and decide on another appropriate flat surface for landing.

Since the Chang'e 2 satellite featured high-imaging precision and a far larger data size than Chang'e 1, scientists and technicians developed mass storage devices for Chang'e 2, with their storage capacity 2.5 times bigger, the transmission speed increased by 80 times and the data transmission capacity enhanced from the original 3 MB to 6-12 MB, providing support to the validation data transmission technology and subsequent lunar exploration missions.

(4) If the Chang'e 2 satellite is compared to a kite, the measurement and control system is the invisible string of the kite. This string served as the only link between the satellite and the Earth. The measurement and control system is composed of the Xichang Satellite Launch Center, Beijing Aerospace Control Center, Xi'an Satellite Control Center, three space tracking ships, six measurement and control stations in China, one overseas measurement and control station, four astronomical observatories and one international networking monitoring station, which form an organic whole.

The Chang'e 2 satellite verified x-band deep space measurement and control technology for the first time, which is the international development trend for this technology. Compared with the increasingly crowded s-band, x-band measurement and control features more abundant measurement and control resources, higher frequency of radio transmission signals, higher orbit measurement precision and easier realization of miniaturization of onboard measurement and control equipment that will result in reduction of weight and enable the space vehicle to have a larger range. X-band measurement and control technology that was verified in the Chang'e 2 mission will play a vital role in future deep space exploration.

▶ Travelling Further into Deep Space

The Chang'e 2 satellite launched on October 1st, 2010, had accomplished all of the six major engineering objectives and the four main scientific exploration tasks set for the mission by the end of its half-year design life on April 1st, 2011. With approval from the overall command system of the program, two important decisions were made after expiration of the design life of Chang'e 2 and its successful

| The Wonderful Beginnings and Progress of the Chang'e Program |

▶ TDI-CCD 3D Camera aboard Chang'e 2

▶ High-speed Data Transmission and Direct Microwave Modulator aboard Chang'e 2

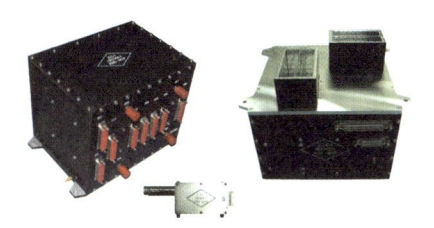

▶ Composition of X-ray Spectrometer System aboard Chang'e 2

▶ Y-Ray Spectrometer aboard Chang'e 2

completion of all engineering objectives and exploration tasks: first, to complete an extended mission, that is, to leave the lunar orbit for the Earth-Sun L2 Lagrangian point (1.5 million km away from the Earth) for scientific exploration; and second, to perform a further extended mission by flying by an asteroid. These two experiments maximized the potential for scientific research and technological experimentation by the satellite.

From 2011 to 2012, the extended experiments of the Chang'e 2 satellite, which surrounded Sun-Earth Lagrange L2 Point for a period of 10 months of scientific exploration, achieved great success. On December 13th, Chang'e 2 reached deep space about 7 million km away from the Earth and brushed past the asteroid Toutatis at a relative velocity of 10.73 km/s. Chang'e 2 came as close as 3.2 km from Toutatis, and imaged the asteroid with onboard monitoring cameras, marking China's first fly-by and exploration of an asteroid.

In just two short years, flying to the Moon 380,000 km away from the

Earth, to the L2 point 1.5 million km from the Earth and further afield to the asteroid 7 million km from the Earth, the Chang'e 2 satellite, carrying the glory and dream of Chinese aerospace workers, kept performing experimental tasks one after another, created and renewed the "height of China" in space, and achieved a series of new breakthroughs in China's lunar and deep space exploration. With success of the extended missions of the Chang'e 2 satellite, China truly realized the engineering objectives of "measuring the orbit precisely, controlling the satellite well and shooting clear photos", and also became the fourth spacefaring power in the world to conduct a successful asteroid mission after the US, the ESA and Japan and also the first country in the world to explore Toutatis. This marked a perfect end to the Chang'e 2 mission.

▶ Diagram on Orbit of Toutatis Developed Based on Photos Shot by Chang'e 2 Satellite

First Landing on the Moon by Chang'e 3

On December 2nd, 2013, China successfully sent the Chang'e 3 lunar probe into space on an improved version of an LM-3B carrier rocket from the Xichang Satellite Launch Center, completing the most important step in lunar landing exploration as part of the second phase of

| The Wonderful Beginnings and Progress of the Chang'e Program |

China's lunar exploration program. Meeting the expectations of Chinese people, Chang'e 3 lunar probe succeeded in soft landing on the lunar surface on December 14th, carrying out landing and patrol exploration by China on a celestial body other than the Earth. The Chang'e 3 lunar probe also served as the first probe to soft-land on the Moon after the Apollo program of the United States, and made China the third country in the world to master landing exploration technologies.

▶ Lunar Lander and "Yutu" Lunar Rover

The Chang'e 3 probe consists of a lunar lander and a lunar patrol vehicle and weighs a total of 3.8 tons. The lunar lander is the most complicated space vehicle ever seen in China, with a total weight of 1,080 kg and a designed service life of 12 months. It carries on board four kinds of scientific equipment: topographic cameras, landing cameras, extreme ultraviolet cameras and lunar-based optical telescopes. With a landing buffer mechanism equipped with four legs that can stretch out and draw back flexibly over a large range, the lander can select a landing point in an automatic and intelligent manner and realize precision hovering and landing.

The lunar patrol vehicle is also called a lunar rover, and it has another pleasant name, "Yutu" (yutu literally meaning "Jade Rabbit", named after the mythological rabbit that lives on the Moon as a pet of the Moon goddess Chang'e). The "Yutu" lunar rover weighs 140 kg and measures 1.5 m in length, 1 m in width

▶ Diagram of Chang'e 3 Lander at Night

and 1.1 m in height. It is designed with a service life of three months, that is, three "days" on the moon. With solar energy as its energy source, it can endure the extreme environment on the lunar surface including vacuum, strong radiation and extreme temperatures, and it carries on board panoramic cameras, lunar exploration radars, infrared imaging spectrometers, particle induced X-ray emission spectrometry, etc. Before Chang'e 3 landed on the Moon, the "Yutu" "hid" on the top of the lander, and "ran" out at the appropriate moment after the landing of Chang'e 3.

▶ Braving the Landing

The Chang'e 3 probe had to first achieve a breakthrough in landing. Chang'e 3 located its landing point in the Sinus Iridum, which is situated at a 43 degrees north latitude and a 31 degrees west longitude and extends about 100 km from south to north and about 300 km from east to west. Its name is derived from the Latin language and means "the Bay of Rainbows". This bay is the most beautiful feature on the moon. As a matter of fact, the Sinus Iridum is a huge meteorite crater, from which lava flowed, flooding a large part of the surrounding area, to form a plain of lava with flat, open terrain and serving a landing point with a high margin of safety. In addition, the area is also typical on the lunar surface with a complicated geological structure and high scientific research value. Meanwhile, the area also enjoys relatively adequate sunshine, and can facilitate ground communications, monitoring and control. More importantly, no other country in the world has explored the Sinus Iridum so far.

On December 2nd, after 19 minutes' flight, the LM-3B carrier rocket sent Chang'e 3 directly into the Earth-Moon transfer orbit with a perigee height of 210 km and an apogee height of 368,000 km. Compared with the seven-day space travel of Chang'e 1, Chang'e 3 travelled much faster, and its orbit injection accuracy was improved by over three times compared with that of Chang'e 2. On December 6th, on instruction from the Beijing Aerospace Control Center, the Chang'e 3 lunar probe had its variable thrust engine ignited successfully and its "space brake" employed smoothly, and after that, it entered into lunar orbit with an average height of about 100 km from the lunar surface and became a true lunar satellite. On December 14th, the variable thrust engine started up once again, and Chang'e 3, which flew around the Moon at a speed of 1.7 km/second, was subject to powered lowering when it was 15 km away from the lunar surface, before finally achieving a perfect landing. On December 15th, the "Yutu" lunar rover began to move toward the transfer mechanism, and after about one hour's careful "exploration", "Yutu" came to stand in front of the transfer mechanism; at this moment, the two "ladders" which supported the "Yutu" gently touched the lunar surface, building a bridge between the probe and the lunar surface. After that, "Yutu" took deliberate steps down the ladders, set foot on the Moon and left two lines behind it. This was the first trail of footprints that a Chinese probe has ever left on a celestial body other than the Earth.

Although the soft landing of Chang'e 3 only lasted a dozen minutes, it could determine the success of the Chang'e 3 mission as a whole. During the landing process, the probe had to first hover stably at a given height in order to observe the lunar surface conditions of the landing point. If the lunar surface conditions are not appropriate for landing, the probe must make modifications by finding another smooth landing point. This process is termed "hovering obstacle avoidance". In this process, all instruments on Chang'e 3 functioned correctly: microwave distance measurement sensors and laser distance measurement sensors measured the distance to the lunar surface and the speed of the probe, landing cameras and optical imaging sensors performed rough obstacle avoidance when approaching the lunar surface, and laser 3D imaging sensors carried out fine obstacle avoidance during the hovering stage, to ensure self-contained guidance, navigation and control of the probe in the landing stage and realize a safe soft landing.

▶ Scientific Exploration on the Lunar Surface

While carrying out scientific exploration, the "Yutu" lunar rover also moved around the lunar lander, so the two could take photos of each other. On the dusty ash-black gravel lunar surface, the sunshine gave the lander a golden hue tinge and the five-starred red flag on the "chest" of the lunar rover was attractive and brightly-colored... There were beautiful photos that the lunar lander and the lunar rover took of each other. Successful imaging of each other, by the lunar lander and the lunar rover, meant smooth operation of all payloads on board these two vehicles, and marked the realization of the objective of this mission, namely, "realizing soft landing and carrying out in-situ exploration and patrol exploration". In the subsequent three months and for even longer afterwards, the lunar rover and the lunar lander will carry out even more scientific explorations.

After landing on the moon, the Chang'e 3 lunar probe faced the problem of how to "survive" despite extreme differences in temperature between day and night. One day on the Moon equals to 27 days on the Earth, with daytime temperatures of up to 150 degrees Celsius and the nighttime temperatures dropping to minus 180 degrees Celsius. To solve the survival problem at night, "Yutu" lunar rover adopted the isotope heat source for the first time. When the night comes, the lunar rover powers off and enters a state of "dormancy" with most equipment on board shut down or on standby. The extended solar wings are also folded to cover the lunar rover just like a quilt. The isotope can keep the instruments on the lunar rover above 20 degrees Celsius below zero. When the sun shines on the lunar surface, "Yutu" wakes up by itself, extends its solar wings once again and continues to work. In daytime, the "Yutu" lunar rover also needs to adjust the angle of its solar wings to avoid being heated too much by the sun. At the hottest point during the middle of the day, "Yutu" can also have a "noon break".

Chang'e 3 lunar probe's first soft landing on a celestial body other than the Earth can enable China to master key technologies in deep space exploration and space science and technology, acquire scientific data from the Moon and the space between the Earth and the Moon, achieve a batch of original scientific results, promote the emergence and development of many new frontier disciplines and interdisciplinary subjects, build a relatively perfect lunar exploration program system and relevant infrastructures, enhance China's capability to integrate deep space exploration systems, realize a giant step forward in the development of space technology and lay a solid foundation for future "sampling and returning".

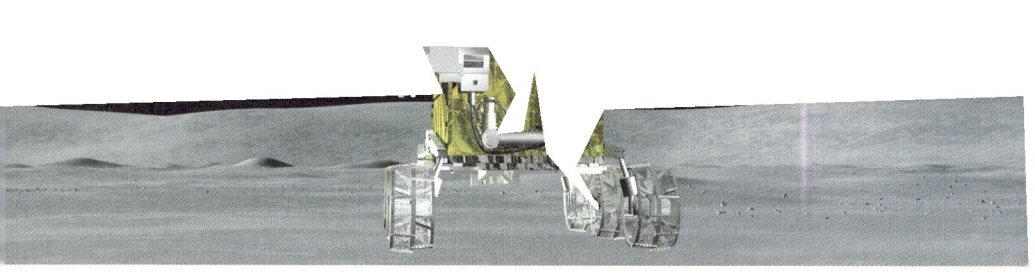

▶ "Yutu" Lunar Rover

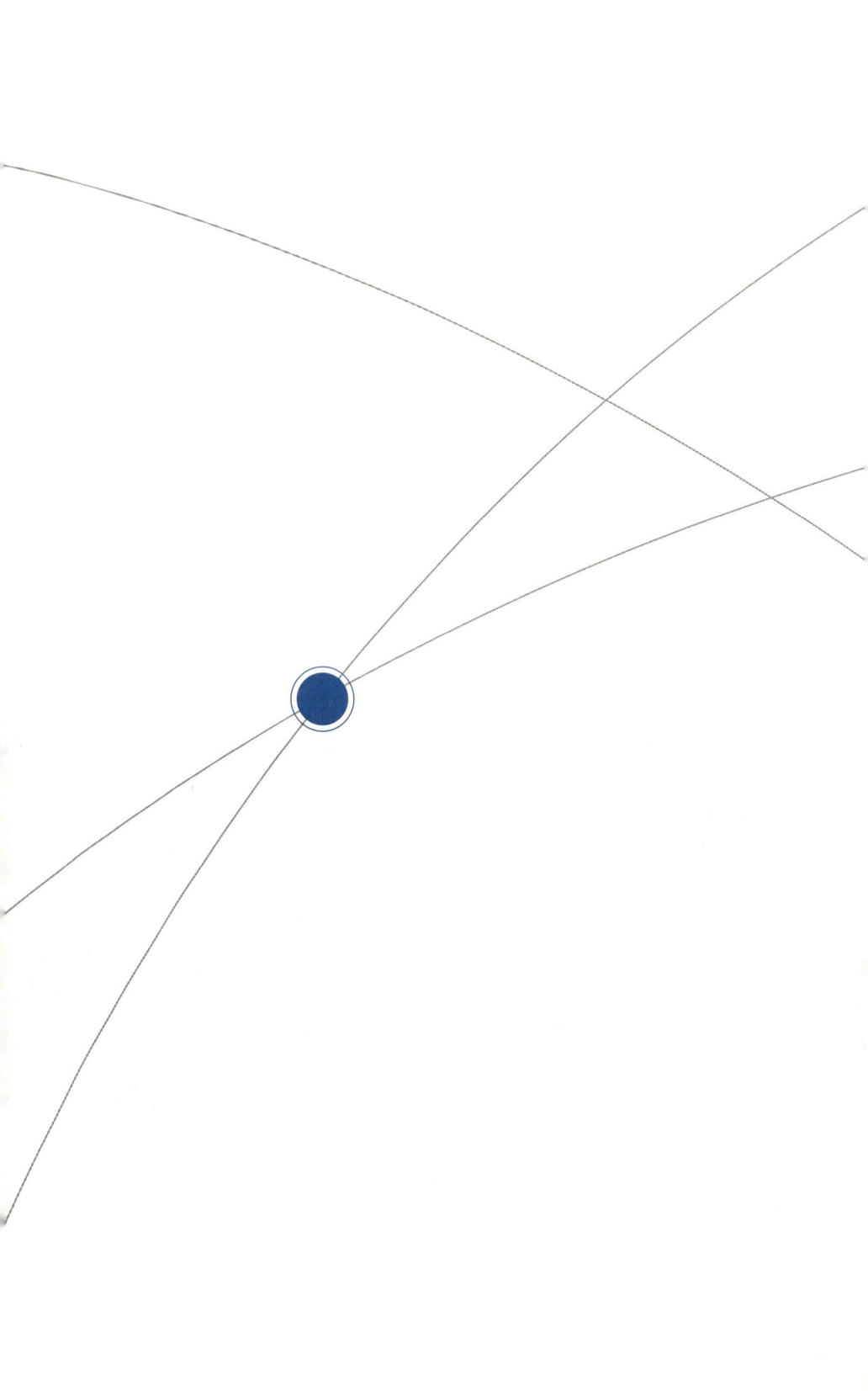

7

Desert, Mountain and Ocean: Spaceports for China's Satellite Launch

As early as 1958, before the launch of China's first man-made satellite, China began to construct space vehicle launch centers. The following decades saw the construction and operation of the Jiuquan Satellite Launch Center, Taiyuan Satellite Launch Center and Xichang Satellite Launch Center in succession. Currently, the Hainan Wenchang Satellite Launch Center is under construction.

First Spaceport for Manned Spaceflight: Jiuquan Satellite Launch Center

The Jiuquan Satellite Launch Center was China's first and also the largest space vehicle launch center. It is geographically located within Ejina Banner of Inner Mongolia and covers an area of about 2,800 square km. With an average altitude of 1,100 m, the environment features flat terrain, a wide field of vision and a dry climate. At the end of the 19th century, when the Swedish adventurer Sven Hadin set foot on the edge of the desert, he lamented with horror: "This is a place no living creature could survive, and it is a sea of death, an awful sea of death."

▶ Difficult Beginnings

When Chinese scientists proposed the development of China's first man-made earth satellite, an exploration team made up of General Chen Xilian, Commander of the PLA Artillery Corps, and Lieutenant General Wang Shangrong, Minister of the General Staff Operations, among others, conducted challenging air and land investigations of seven preselected areas in Northeast China, North China and West China at the beginning of 1958 and finally chose the Gobi desert on the north of Jiuquan as China's first missile and satellite test and launch site. For this, the team developed an investigation and site selection report and submitted it to the central authorities.

In April, the Military Commission of the CCCPC resolved to organize special forces for the construction of the missile and atomic bomb test range. For this purpose, the 19th regiment, the headquarters of the engineer corps of the voluntary army, the second division of the logistics department of the voluntary army and other departments stationed in North Korea were brought home. After strict political examination, the "PLA Special Engineering Headquarters" was set up in Beijing and codenamed 7169.

In the desolate and uninhabited Gobi desert, tracts of green military tents mushroomed overnight, and 100,000 officers and soldiers stationed themselves here, beginning China's journey from desert to space.

At that time, the space force was faced with harsh natural and living conditions in the desert. The vast Gobi desert was tough and could make you crazy. A gust of wind could turn the sky dark, lift the heavy compartment of the train off the rail and blow the large barrels filled with oil several kilometers away, not to mention the cars, people and supplies. Some personnel went out to work and encountered strong wind. As they had nothing to hold on to, they were swept away by the wind forever. The tenth division of the railway corps, responsible for construction of the railway from Qingshui to the launch site which extended to more than 270 km, were distributed along the desert line over several hundred kilometers. When they encountered strong winds, they had to lie with their stomach to the rails while holding onto the rail to avoid being blown away by the wind.

The special engineering headquarters included a general, five majors and more than 100 colonels and senior colonels. They all lived in individual mud huts. These rooms served both as their sleeping quarters and their offices. Hard as the conditions were, it was very convenient.

China's Journey to Space: From Dream to Reality

They started work straight after leaving the covers and went to bed straight after getting under the covers. They seemed to have gone back to the war years, and even the celebrated General Chen Shiju was no exception. With their hard-working spirit, these old generals set an example for the 100,000 officers and soldiers and encouraged them to have courage and confidence to overcome difficulties. It took them just two years and six months to complete the first phase of the project.

This heroic rocket troop weathered the twelve long and tough years from 1958 to 1970. More importantly, they turned the "sea of death" into China's first "spaceport" to space with their wisdom and hard work and witnessed the incredible development of China's aerospace cause. On

▶ Aerial View of Jiuquan Satellite Launch Center

September 10th, 1960, China successfully launched its first short-range missile with domestic fuel from here; on October 27th, 1966, China's first nuclear missile obtained lift-off from here; and on April 24th, 1970, China's first man-made earth satellite was also shot into space from here.

▶ Modern Manned Spaceflight Launch Center

In 1992, the initiation of China's manned spaceflight program brought the Jiuquan Satellite Launch Center new development opportunities. However, with limited funding, how they were going to build an internationally advanced launch site using the "three vertical" model became the focus of the demonstration of aerospace experts.

At that time, China had, for several decades, used the traditional horizontal assembly and test mode in the technical sector with relatively established technologies and lower construction costs. The shift to the "three-vertical" launch mode necessitated the adoption of a new design and technological reconstruction. In foreign countries, a similar launch site generally cost more than USD one billion, which almost equaled to China's total investment in the entire manned spaceflight program. China's investment in launch site construction was only RMB one billion.

While the launch site construction proposals were involved in a new round of demonstrations, the proposal for the horizontal tests used in the technical sector took over, and the "three vertical" proposal was on the verge of abortion. Qian Xuesen, a strong advocate of China's manned spaceflight program, had been paying attention to the latest developments of the program. When he learnt that the "three vertical" proposal failed to pass the demonstration, he got worried and wrote to the office of the COSTIND. He pointed out in the letter that the vertical mode and long-distance launch control was the trend in the international aerospace industry so China should employ this in order to construct an internationally advanced manned spaceflight launch center. The opinion of Qian Xuesen steered the direction of the launch site proposal demonstration. Finally, the "three vertical" proposal was passed. On July 3rd, 1994, China's manned spaceflight launch center officially opened.

After four years of effort, buildings of different shapes and sizes sprang up one after another and a world-class space launch center miraculously stood before the world. The Jiuquan Satellite Launch Center was once again under the spotlight. From the Shenzhou 1 experimental spacecraft's travel to space in 1999 to the Shenzhou 10 spacecraft putting three astronauts into space in 2013, the Jiuquan Satellite Launch Center witnesseds milestones one after another in China's manned spaceflight program.

The Jiuquan Satellite Launch Center is composed of the technical zone, the launch zone, the test command zone, the test coordination zone and the astronaut zone. The technical zone provides the site for technological preparations before rocket and spacecraft launch. After the rockets, spacecraft, astronauts, payloads and escape tower

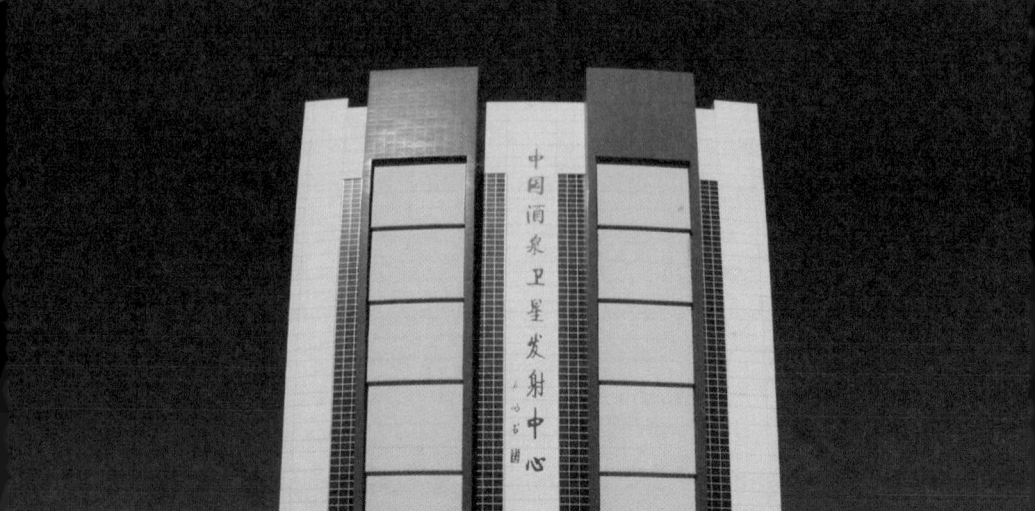

▶ Jiuquan Satellite Launch Center

Desert, Mountain and Ocean: Spaceports for China's Satellite Launch

enter the launch site, they first undergo testing, assembly and various examinations here.

The most eye-catching facility in the technical zone is the vertical assembly and test plant. It is a reinforced concrete structure 100 m high. With its height equal to a building of 38 stories, it has become the largest single story building in Asia. The vertical assembly and test plant covers an area of more than 30,000 square meters and is China's first and also Asia's largest vertical assembly and test plant. The plant is large in scale and, more importantly, it is a huge framework structure with no floorslab throughout the structure, which is unprecedented in China and even Asia's architectural history and has been hailed by foreign experts as a "Chinese-style miracle".

▶ Ask-the-Sky Pavilion in Jiuquan

The launch zone is responsible for preparations before rocket and spacecraft launch and also for their actual launch. It is situated about 1,500 km to the southeast of the technical zone and is simple in layout. Although there are few structures above ground, the world under it resembles a labyrinth, with underground equipment rooms, astronaut safe shelter rooms, flame diversion troughs, etc.

The landmark architecture in the launch site is certainly the launch tower, a steel structure fixed tower. It measures more than 100 m in height and weighs 2,500 tons, with its largest cantilever as long as 24 m. If a rocket wants to fulfill its dream of flying high, it needs to leave the embrace of the launch tower. If the launch tower is compared to a "mother", then the wires on the launch tower are the "umbilical cord" connecting the "mother" and her "baby". These wires will release from the rocket until the moment immediately before the launch of the rocket.

For this reason, the launch tower is also termed the "umbilical tower".

The greatest difference between a manned spaceflight launch tower and a satellite launch tower is that there are astronauts on board a spacecraft. If any accidents occur with the rocket or the spacecraft, the astronauts need to evacuate to a safe area in time. Therefore, the manned spaceflight launch tower is designed with high-speed explosion-proof elevators, emergency evacuation slides and other safety facilities. If anything unexpected occurs, astronauts can get to the safe shelter rooms under the launch tower using these facilities, thereby safely avoiding any threat to their lives.

Between the vertical assembly and test plant and the launch tower, there is a special railway 1,500 m long and 20 m wide, which can be claimed to be the widest railway in China. This railway is mainly used to transport the movable platform carrying the spacecraft and rocket complex from the vertical assembly and test plant to the launch tower.

The astronaut zone is the training and living place for astronauts

▶ Launch Command Center in Jiuquan

after they arrive at the launch site. It is also an innovative feature separating the manned spaceflight launch center from other launch sites. More generally, it provides astronauts with relevant training, medical monitoring and care and living support. In the astronaut zone, the red walls and white tiles are set against the clear sky and are surrounded by green trees and the Populus euphratica forest on the bank of the Ruoshui River. All combine to complement each other's beauty, making the whole astronaut zone exquisite and elegant. It also has a lovely name: the "Dream Realization Garden". The most mysterious, and also the most crucial, part of the Dream Realization Garden is the "Ask-the-Sky Pavilion". Situated in the southwestern corner of the test command zone of the launch site, it is a white-and-jacinth architectural complex. The pavilion combines natural scenery and artificial landscape, and interior decoration and outdoor buildings in an organic manner, creating an oasis in the desert for the astronauts.

After the hard work of several generations of Chinese people, China finally turned the isolated Gobi desert into a modern launch site using their self-reliance, diligence, wisdom and will. Nowadays, the Jiuquan Satellite Launch Center is connected to the inland via railways and highroads. In addition, a nonstop air route to Beijing and a developed communications network have been put into service. These three "dimensions" of transportation are bringing the small Gobi town increasingly closer inland, with scientific and technological information and living resources constantly flowing into the launch center.

China's Journey to Space: From Dream to Reality

Aerospace Town on the Plateau in Northwestern Shanxi: Taiyuan Satellite Launch Center

▶ Taiyuan Satellite Launch Center

The Taiyuan Satellite Launch Center sits on the plateau in the northwest of Taiyuan City, Shanxi Province, lying in the temperate zone and neighboring the Luyashan attraction. The climate here features a long winter, no summer and autumn immediately following spring, and it has an annual frost-free period of 90 days and an annual average temperature of 5 degrees Celsius.

The construction of the Taiyuan Satellite Launch Center dates back to 1967. At the beginning of March of that year, there was still a chill in the air from the early spring. A troop of more than 1,000 members took a military boxcar from the Jiuquan Satellite Launch Center and arrived at Kelan County, Ningwu, Shanxi Province after four days of travel. The reason for selection of Kelan County as the construction site for the space vehicle launch center was attributed to its important geographical location: it occupies Guanningwu on its east, faces the surging Yellow River on its west, backs the magnificent Great Wall on its north and borders the ancient Town of Jinyang on its south.

As the construction of the ground facilities for missile launch was to be completed before 1968, the engineering tasks for the first phase of the Taiyuan Satellite Launch Center project were very urgent. To make matters worse, there was no flat road in Kelan County, let alone a railway; there was no power supply, not even a sufficient water supply available for the troops; and more than 20,000 villagers in a dozen villages needed to be relocated. In face of these obstacles, the troops first solved the problems of road construction, power connection, well-digging and villager relocation. To speed up progress on the project, more than 3,000 new soldiers were recruited from various regions and 28 university graduates were assigned to work there, ensuring completion of the first phase of the project on schedule.

On December 18th, 1968, despite the low temperature of 36 degrees Celsius below zero, China's first self-developed medium-range liquid-fuelled carrier rocket was successfully launched. In January 1976, the Taiyuan Satellite Launch Center was officially founded. In September 1988, it sent China's first meteorological satellite, FY-1A, on an LM-4 carrier rocket into sun-synchronous orbit. In May 2002, it also succeeded in boosting China's first ocean satellite, HY-1A. To date, it has created nine first achievements in China's satellite launch history.

Since the entrance of China's aerospace and launch technologies into the international market, the Taiyuan Satellite Launch Center has undertaken the launch of many foreign commercial satellites. In the 1990s, the Taiyuan Satellite Launch Center signed the contract on "Iridium" communications satellite launch services with Motorola Satellite Communications Group in the US. Subsequently, since September 1997, the Taiyuan Satellite Launch Center has performed six consecutive launches by means of launching "two satellites on a single rocket" and successfully sent 12 Iridium satellites into preset orbit. In the networking launch, the center won high praise of the client for its unsurpassed accurate launch time and orbital injection accuracy. After that, the Taiyuan Satellite Launch Center also succeeded in sending four commercial satellites from Brazil, the European Union and other countries and regions into space. On October 14th, 1999, it also shot China's first resource satellite into sun-synchronous orbit. Successful launch of the satellite that was hailed as a "model of south-south high technology cooperation" further strengthened the international influence of the Taiyuan Satellite Launch Center.

"Moon City" Helping Chang'e with the Journey to the Moon: Xichang Satellite Launch Center

More than 60 km's travel from Xichang of Sichuan Province will bring you to the mysterious Shaba valley. A stream flows from the high Maoniu Mountain down into the valley, jingling and meandering. The valley is flanked by high mountains with lush tsugae mushrooms

| Desert, Mountain and Ocean: Spaceports for China's Satellite Launch |

and picturesque farmland, and from the farmhouses situated at the foot of the mountains and beside the stream rise thin curls of smoke. All these factors conjure up the image of "a retreat away from the world" (an imaginary ideal world in the literary works "The Story of the Peach Blossom Valley" by Tao Yuanming, a poet of the Eastern Jin Dynasty). However, who would have thought that a launch site would spring up from this deep and serene valley? State-of-the-art scientific and technological equipment is in a unique and sharp contrast

▶ Xichang Satellite Launch Center

to the traditional Yi villages. Nowadays, the Xichang Satellite Launch Center has become the launch site for the "lunar exploration program", capturing attention from around the world.

In terms of the geographical location, the Xichang Satellite Launch Center enjoys many advantages. First, it has a high altitude and a low latitude and the latitude of the launch site has something to do with the inclination of the satellite orbit. It is located at a longitude of 102 degrees east and a latitude of 28.2 degrees north with an average elevation of 1,500 m. The lower latitude means a closer distance to the equator, full utilization of the centrifugal force as a result of rotation of the Earth, a shortening of the distance from the Earth to the satellite orbit, the saving of rocket fuel and finally an increase in rocket payload.

Second, the location has an unremarkable terrain and a solid geological structure. Some of the basins which formed in the valley extending from the south to the north not only benefit the overall layout of the launch site and the deployment of ground launch facilities, technological equipment and tracking, measurement and communications networks, but also satisfy future requirements in relation to expansion and further development.

Third, it enjoys a favorable climate. Xichang is located in the subtropical plateau monsoon climate zone with an annual average temperature of 16 degrees Celsius. It is one of the regions which has the smallest temperature changes in China. For this, it has also earned the reputation as the "junior spring city", where "there are blooming flowers in various color throughout four seasons, and it is like spring all year round with a warm winter and a cool summer." In terms of its climate, Xichang features distinct rainy and dry seasons, as many as 320 sunny days annually, almost no foggy days and a low annual average wind speed. These advantageous climatic conditions greatly increase the annual test cycle and launch windows. Ancient people once used these poetic lines to describe the bright moonlight in Xichang: "The reflection of the bright moon moves in the water of Qiongchi Lake, where the water is so clear that it can even reflect all that in the sky". This has earned Xichang the lovely name of the "Moon City".

Fourth, Xichang has sufficient and stable water resources and thus can meet the water demand of the launch site for waste cleaning and equipment cooling.

| Desert, Mountain and Ocean: Spaceports for China's Satellite Launch |

Fifth, its transportation and communications conditions are also ideal. The launch site is not far away from Xichang Qingshan Airport, the Chengdu-Kunming Railway and the Sichuan-Yunnan Highway. Moreover, the Jinsha River waterway on its east leads to Yibin, Chongqing and even Shanghai. These conditions facilitate transportation of necessary materials and satellite and rocket products. Furthermore, the national communication trunk is also adjacent to the launch site and able to ensure communications during launch tests. In summary, the Xichang Satellite Launch Center is richly endowed by nature and ideally located.

In September 1970, the Jiuquan Satellite Launch Center manoeuvered veteran soldiers and able commanders and pooled various forces nationwide to Xichang. At the time, the Chengdu-Kunming Railway which went through many tunnels and over many bridges had just laid

▶ Sculpture for Xichang as Moon City

its tracks and no passenger or cargo train had yet run on this railway. The construction forces then boarded the trains and winded along the railway tracks. The large-scale transportation lasted more than two months. With instruction from Premier Zhou Enlai, the Xichang Satellite Launch Center was incorporated in the Fourth Five-Year Plan of the state, which accelerated the pace of the construction.

At the end of 1980, the construction of the infrastructures of the Xichang Satellite Launch Center was completed, which was then officially delivered for use in 1982 with the LM-3 carrier rocket standing in the previously empty valley. On April 8th, 1984, China's first geostationary experimental communications satellite DFH-2 was successfully launched from the Xichang Satellite Launch Center. In February 1986, the center also sent China's first operational communications satellite into preset orbit, putting an end to the previous practice of Chinese people renting foreign satellites to enjoy their television and radios. The Xichang Satellite Launch Center then began to get the attention of the world.

In 1986, the Xichang Satellite Launch Center opened its arms to the world and demonstrated its high aspirations to travel to space by undertaking its foreign commercial satellite launch business. As China's first space vehicle launch center that had opened to the world, it attracted close attention and strong interest from many countries and companies, which then visited Xichang for investigations. On October 8th, 1988, US Defense Secretary Weinberg visited the Xichang Satellite Launch Center and said to the reporters present: "This launch site meets satellite launch conditions and has great potential. I am also deeply impressed with its further improvements of the facilities so as to carry out China's own space program and launch foreign commercial satellites."

On April 7th, 1990, the Xichang Satellite Launch Center also saw LM-3 succeed in lifting off the USA-made AsiaSat 1 into predetermined orbit, making China the world's third nation and also Asia's first country to foray into the international commercial launch market. Also from the Xichang Satellite Launch Center, Optus B1 and B3 satellites developed by the US, Apstar satellite, AsiaSat 2, Mabuhay satellite and other international communications satellites were also placed into space. With this series of satellite launches, China became a model of participation in international aerospace technology exchanges and cooperation. Since the launch of its first commercial satellites, the Xichang Satellite Launch Center has sent about 20 Chinese and foreign commercial satellites into space serving as China's modern launch site with the longest history, the most launches and the highest degree of openness.

In 2004, with the initiation of China's lunar exploration program, the Xichang Satellite Launch Center also become a "lunar exploration spaceport" in China's aerospace sector. On October 24th, 2007, Chang'e 1 satellite carried out a "zero window" launch at the Xichang Satellite Launch Center, saved 120 kg of satellite fuel and prolonged the on-orbit operation duration by four months, thus composing a new chapter in China's aerospace history on the Moon 380,000 km away from the Earth. In October 2010, the Chang'e 2 satellite once again perfectly accomplished a "zero window" launch and flew directly to the lunar trajectory, laying a solid foundation for "landing" of the Chang'e 3 satellite onto the Moon scheduled in 2013.

The Xichang Satellite Launch Center has made great strides in its developments with regard to six aspects: from the launch of single-type rockets to the launch of multi-type rockets, from single-direction launch to multi-direction launch, from the launch of geostationary satellites to the launch of multi-orbit spacecraft, from the launch of Chinese satellites to the launch of international commercial satellites, from three satellite launches at most to ten satellite launches annually, and from close-control test launch to remote-control launch. With these achievements, the Xichang Satellite Launch Center has stepped into the league of the world's top ten space vehicle launch centers.

Future Manned Lunar Landing Spaceport: Wenchang Satellite Launch Center

The Hainan Wenchang Satellite Launch Center is China's fourth launch site under development and located in Longlou Town, Wenchang City, Hainan Province. To be in line with the sustainable development strategy of China's aerospace sector, the construction of the Wenchang Satellite Launch Center was formally opened in September 2009.

The Wenchang Satellite Launch Center is the launch site with the lowest latitude and is mainly designed to launch the new heavy lift LM-5 series of rockets. The Wenchang Satellite Launch Center has never aimed to be a reproduction and repetition of existing launch centers. Instead, it aims to become a new world-class modern space vehicle center featuring strong comprehensive launch capabilities, high safety, reliability and IT-application levels and ecological and environmental protection, by making more advanced innovations and developments and applying the most advanced design concept with the latest technologies in the world's aerospace sector. With a total area of 40 square kilometers, the Wenchang Satellite Launch Center consists of a space vehicle launch port, a space science theme park, a rocket assembly plant, a rocket launch base, a command center, etc.

In future, the launch missions of China's

▶ Planning Drawing of Wenchang Satellite Launch Center

| Desert, Mountain and Ocean: Spaceports for China's Satellite Launch |

▶ Groundbreaking Ceremony of Wenchang Satellite Launch Center

manned spaceflight, deep space exploration and manned lunar landing programs will all be implemented at the Wenchang Satellite Launch Center.

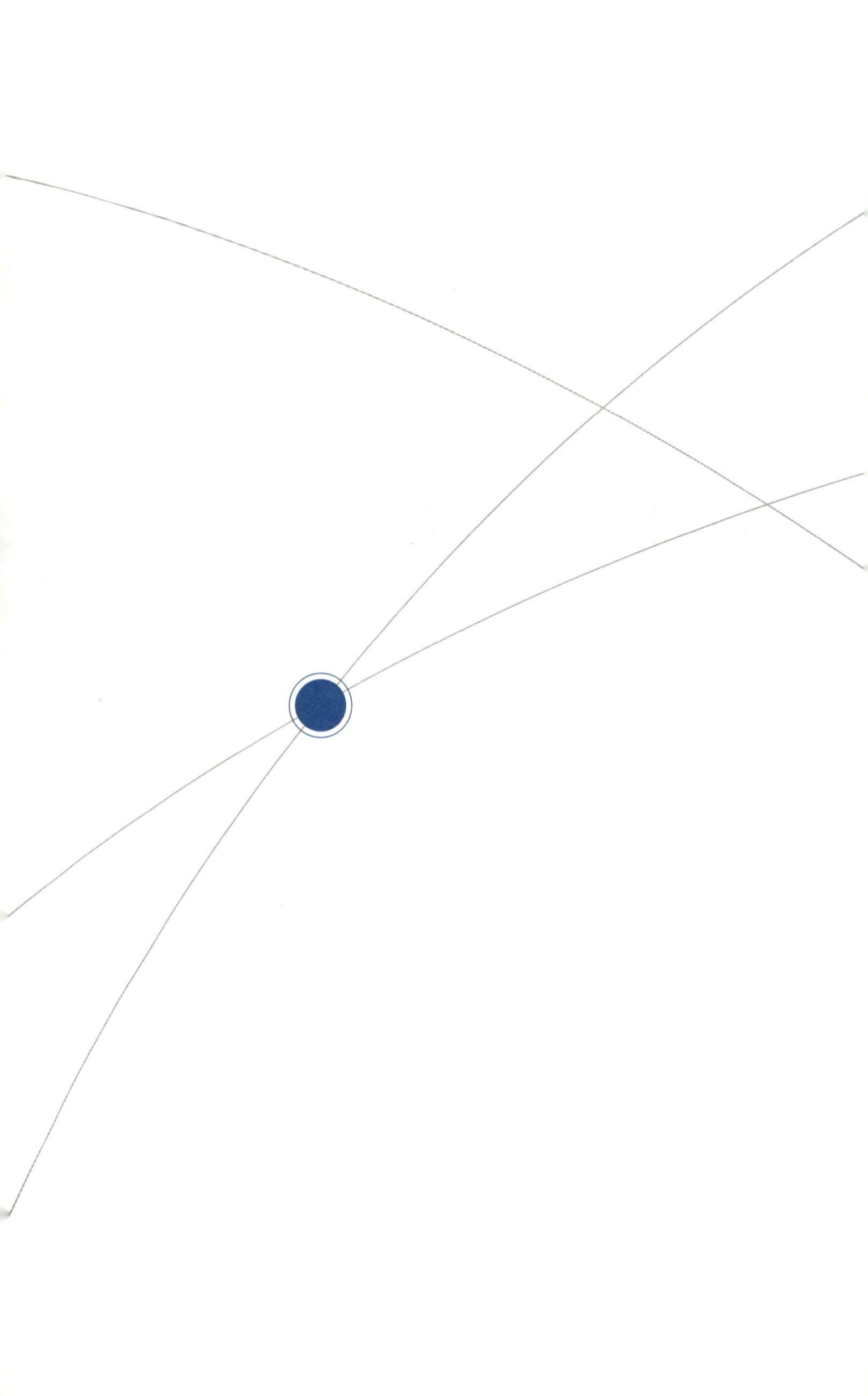

8

The Broad Prospects of China's Aerospace Sector

China's Journey to Space: From Dream to Reality

Since the beginning of the 21st century, China has implemented a broad range of significant systems and special programs, including the manned spaceflight program, the lunar exploration program, the Beidou satellite navigation system, the large capacity communications satellite program and the high-resolution earth observation system and ranked among the world's most advanced countries in some cutting-edge technology fields. Looking into the future, China will strengthen basic capability building of the aerospace industry, deploy leading technology research in a prospective manner and drive comprehensive, coordinated and sustainable development of the aerospace cause.

New Carrier Rockets

China will step up construction of its space transportation system by keeping improving the carrier rocket type spectrum, enhancing their spaceflight capabilities and carrying out maiden flights of the LM-5, LM-6 and LM-7 carrier rockets.

The LM-5 series of carrier rockets are China's newly developed large-scale cryogenic carrier rockets involving a large technological span and many key technological breakthroughs. In addition, their carrying capacity has also been substantially increased. For example, their carrying capacity to near-Earth orbits has been raised

▶ LM-5 Carrier Rocket

from 9.5 tons to 25 tons. Therefore, LM-5 marks great jump forward in the development of Chinese carrier rockets in terms of their carrying capacity.

▶ Diagram of LM-6 Rocket

The LM-7 carrier rockets represent a new generation of highly reliable and safe medium launch vehicles to meet the demand of China's manned space station program for the launch of cargo spacecraft. They are characterized by high reliability, non-toxic propellant and no pollution.

Various Artificial Earth Satellites

China will focus on the construction of a space infrastructure framework made up of earth observation, communications and broadcast and navigation and positioning satellites, so as to facilitate the initial formation of long-term, continuous and stable business service capabilities and develop new scientific and technological experiment satellites

▶ Earth Observation Satellites

China will perfect current "Fengyun", "Ziyuan" and "Haiyang" satellite families and the "environment and disaster monitoring and forecasting satellite constellation", and develop a new generation of meteorological satellites, three-dimensional mapping satellites, environment and disaster monitoring and forecasting satellites and other new earth observation satellites, with the view of forming an all-weather, around-the-clock, multi-spectral and stable earth observation system with various resolution options.

▶ Communications and Broadcast Satellites

China will better its fixed service satellites, television and broadcast satellites and mobile service satellites and develop a new generation of ultra-large satellite platforms with larger capacity and higher power, namely, DFH-5.

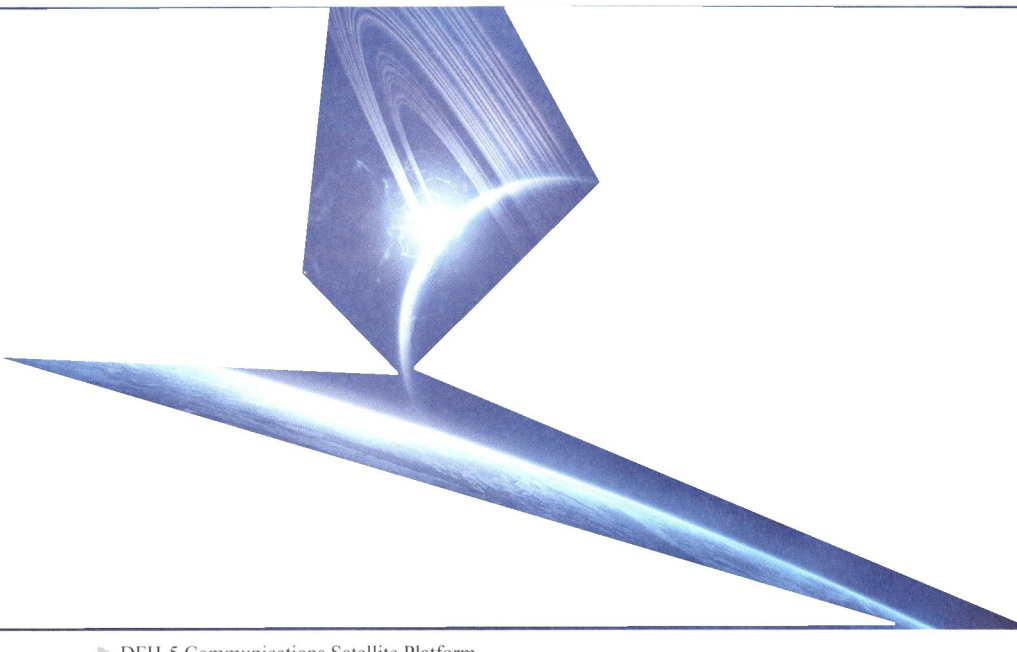

▶ DFH-5 Communications Satellite Platform

▶ Navigation and Positioning Satellites

In order to construct China's self-developed and independently-operated satellite navigation system with global coverage that is also compatible with other satellite navigation systems in the world by 2020, China will develop several navigation satellites based on the Beidou regional satellite navigation system built at the end of 2012.

▶ Scientific and Technological Experiment Satellites

China will develop satellites dedicated to launching hard x-ray modulation telescopes and initiate quantum science experiment satellite and dark matter exploration satellite programs.

Future Manned Space Station

▶ Concept Drawing of Tiangong 2 Space Lab

In 2015, China will launch the Tiangong 2 space lab, China's second self-developed space lab after Tiangong 1. It will further validate space rendezvous and docking technology, perform earth observation and carry out applications and experiments in space and earth system science, new technology for space applications, space technology and aerospace medicine. Upon the launch of the Tiangong 2 space lab, an accompanying miniaturized satellite will be released and docked with the cargo spacecraft later.

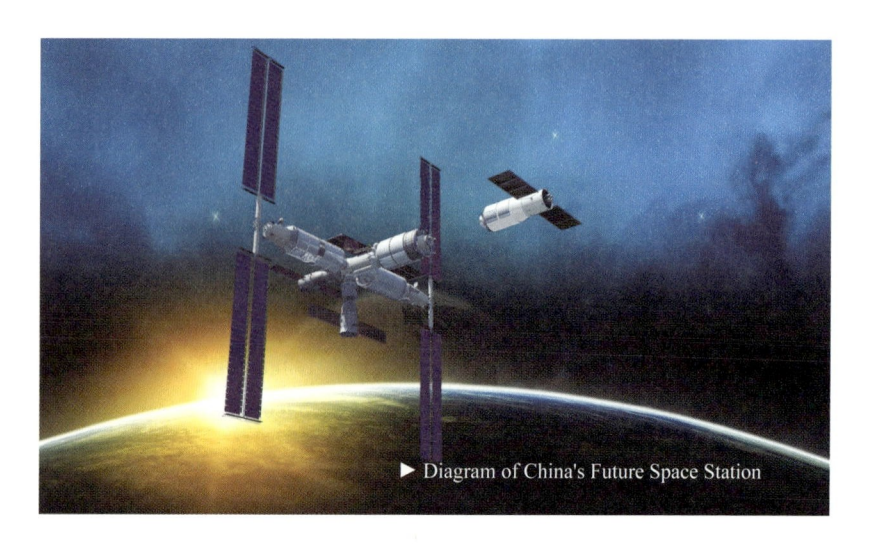

▶ Diagram of China's Future Space Station

China's future space station will adopt a T shape and consist of three 22-ton modules with the core module in the middle and Experiment Module I and Experiment Module II located at the left and right side of the core module respectively. The space station will be equipped with an airlock module above it to facilitate the process of astronauts leaving the module and it will also be fitted with two mechanical arms to assist in docking, material supply, extravehicular operation and scientific experiments. During operation of the space station, a second core module will be launched and docked with the space station. In the end, China's space station will be a cross shape allowing for further module expansion.

China's manned space station will serve as an important base for China's research on space science and new technologies and will have an on-orbit service life of more than ten years. In space station technology and space application, China will adopt a more open cooperative attitude and manner in order to increase cooperation and exchanges, actively carry out international cooperation and share space resources.

Deep Space Exploration in Progress

China will continue its lunar exploration program according to the three-phase development ideas of "orbiting, landing and returning", proceed with deep space exploration demonstration and carry forward exploration of planets, asteroids and the Sun in the solar system.

The objective of the second phase of China's lunar exploration program, namely "landing", is to realize soft landing on the lunar surface and independent patrol exploration. The second phase includes the three missions of Chang'e 2, Chang'e 3 and Chang'e 4. Currently, China has successfully launched Chang'e 2 and Chang'e 3, and Chang'e 4 serves a back-up to the Chang'e 3 mission. After Chang'e 3 completes its mission, technicians will make adaptability modifications on Chang'e 4 by optimizing the engineering tasks and the scientific objectives, so as to verify the key technologies needed for Chang'e 5.

The third phase of China's lunar exploration program, that is "returning", is aimed to realize unmanned sampling and returning, and it consists of the Chang'e 5 and Chang'e 6 missions. Chang'e 5 will be the first sample return probe for the "sampling and returning" mission, and it is scheduled to be launched around 2017 and from the Hainan Wenchang Satellite Launch Center. It will carry a lunar lander, a lunar rover, a lunar ascender and a return module on board. It is designed to carry out terrain exploration and geological surveying at the landing area, obtain lunar soil samples, analyze sites related to lunar soil samples and establish contacts between field exploration data and lab analysis data.